SCIENTIFIC PAPERS

S. B. McLaren

SCIENTIFIC PAPERS

MAINLY ON ELECTRODYNAMICS
AND NATURAL RADIATION

INCLUDING THE SUBSTANCE OF
AN ADAMS PRIZE ESSAY IN THE
UNIVERSITY OF CAMBRIDGE

BY THE LATE

SAMUEL BRUCE M^cLAREN, M.A

OF THE UNIVERSITY OF MELBOURNE AND TRINITY COLLEGE CAMBRIDGE
PROFESSOR OF MATHEMATICS IN UNIVERSITY COLLEGE READING

CAMBRIDGE
AT THE UNIVERSITY PRESS
1925

PREPARED FOR PUBLICATION BY

HENRY RONALD HASSÉ, D.Sc.,
Professor of Mathematics in the University
of Bristol:

THOMAS HENRY HAVELOCK, F.R.S.,
Professor of Mathematical Physics in the
Armstrong College, Newcastle-on-Tyne:

JOHN WILLIAM NICHOLSON, F.R.S.,
Fellow and Tutor of Balliol College, Oxford:

JOSEPH LARMOR, Kt., F.R.S.,
Lucasian Professor of Mathematics, Cambridge.

CAMBRIDGE
UNIVERSITY PRESS

University Printing House, Cambridge CB2 8BS, United Kingdom

Published in the United States of America by Cambridge University Press, New York

Cambridge University Press is part of the University of Cambridge.

It furthers the University's mission by disseminating knowledge in the pursuit of
education, learning and research at the highest international levels of excellence.

www.cambridge.org
Information on this title: www.cambridge.org/9781107682771

© Cambridge University Press 1925

First published 1925
First paperback edition 2013

A catalogue record for this publication is available from the British Library

ISBN 978-1-107-68277-1 Paperback

Cambridge University Press has no responsibility for the persistence or accuracy of
URLs for external or third-party internet websites referred to in this publication,
and does not guarantee that any content on such websites is, or will remain, accurate
or appropriate.

PREFACE

IN the *Cambridge University Reporter* under date March 14, 1911, the subject for the biennial Adams Prize Essay to be submitted in December 1912 was announced as follows:

The Theory of Radiation.

The experimental scrutiny of the spectra of gaseous substances has amassed much knowledge, already expressed in semi-empirical formulae, relating to the structure of spectra, whether composed of discrete lines or of bands, and also relating to the influence of various physical causes, such as the admixture of other substances, on the relative intensities of the lines.

The nature of the magnetic influence on lines and bands which is exhibited in the Zeeman effect awaits closer investigation; and the classification of the lines of a spectrum which it suggests may afford further clues towards the assembling of those groups of lines which are possibly in some way components of one fundamental mode of vibration.

A critical discussion of some of the problems of molecular dynamics which are associated with these phenomena is proposed. This discussion might proceed either from some hypothesis of structure of the molecules of matter, or from comparison by way of analogy with the properties of known types of vibrations. Questions relating to the constitution of natural radiant energy in statistical equilibrium might also come up for consideration.

In the award the Prize was divided between an essay submitted by S. B. M⁗Laren and the work of another writer who devoted himself to a different branch of the subject. Neither of these essays was in form ready for immediate publication. The final adaptation of Mr M⁗Laren's essay, which was intended to contain a digest of his previous publications on related topics, was proceeding (cf. p. 8 *infra*) when the Great War intervened; and it had to give way to a higher call of duty. His death from wounds at Abbeville on August 13, 1916, might have made an end of the whole project.

The present volume, the preparation of which is due to the devotion of some of his personal friends, will serve as a tribute to his memory and a memorial of his scientific activity. It contains an adequate account of the Adams Prize work, so far as it has not been superseded by the abundant activities of other investigators in the ten years which have since elapsed: and it is completed by reprints of scientific papers, or by systematic accounts of their contents where reproduction was deemed unnecessary.

Acknowledgment is due to the Editors of the *Philosophical Magazine* and *Nature,* and to the Councils of the London Mathematical Society and British Association, for permission to reprint. Thanks are due to Professor Hugh Walker of Lampeter and Professor H. S. Allen of St Andrews for valuable assistance in the undertaking.

J. L.

November 11, 1924.

CONTENTS

A PERSONAL APPRECIATION

By Prof. HUGH WALKER, M.A., LL.D.

St David's College, Lampeter.

I first made the acquaintance of Bruce McLaren at Birmingham in 1908 or 1909, when he was lecturer in mathematics in the university of that city. There were more than twenty years of life between us, and of the science in which McLaren was primarily interested I knew nothing. Nevertheless, the acquaintance prospered, and I think it is safe to say that in the latter years of his life no man in England knew him so intimately as I. We were both devoted to things of the intellect, and he was a man far too whole and round to allow himself to be utterly absorbed in mathematics alone. That is perhaps the first thing to grasp with regard to Bruce McLaren. His ruling passion was the passion for truth, wherever she was to be found. He was interested in philosophy, in social science, in literature and art as well as in mathematics and physical science. He was interested above all in life, and I think he would have said that the mathematician who was nothing but a mathematician knew little more of the art of living than O. W. Holmes's Scarabee. It is true he had at times the abstract thinker's absorption and absent-mindedness. But he could rouse himself and prove himself very much alive and very much the man all round—far more so than those who knew nothing about the "strange seas of thought" whereon he habitually voyaged.

Though McLaren and I were drawn to one another by a kind of pre-established harmony, to which my total ignorance of and incapacity for mathematics seemed to be no barrier, I don't think the friendship between us would have prospered as it did had it not been helped by my wife. She understood him far sooner than I did, and his position appealed to the mother in her. Of Scottish blood, born in Japan and brought up in Australia, this most affectionate and domestic of men had been

M.

1

drawn by the gift within him as by a magnet to Cambridge; and now he was living alone, with many acquaintances but very few friends. For while he was superficially frank enough, there was an inner guard of reserve, due possibly to his Scottish ancestry, which few ever penetrated. Perhaps it was because I inherited from the same blood the same reserve that in the end I was one of the privileged few who did penetrate it. His solitude could not fail to appeal to a woman who understood the hunger of the heart that it caused. And so presently, without words or forms and almost without his knowledge or mine, but to the great happiness of both of us, he was established as a sort of adopted son. And his father and mother in distant Australia were big enough in soul and secure enough in his inalienable love to sanction it and welcome it. He used to talk of his English mother and his Australian mother. He did not speak so of me; our relation was a peculiar and complex one—half that of father and son, half that of elder brother and younger; but always and altogether that of perfect friends.

It followed, of course, that he was welcome to come when he pleased to his English, or rather his Welsh, home. He knew us first as the inhabitants of a poky little house in Birmingham. But that was our home only for part of the vacations; and he used frequently to come to Lampeter. He was charmed with its beauty, and surprised to find among the hills of Wales a small college which seemed to be a strayed fragment of an English University—Oxford, rather than Cambridge. On his first visit he elected to approach us by a way of his own. He sent on his bag by rail, and tramped alone across one of the most extensive unbroken tracts of moorland south of the Tweed. It is sixteen miles as the crow flies from Rhayader to Strata Florida, and much the greater part of it is roadless and almost if not quite trackless. He arrived on a beautiful summer afternoon, somewhat dusty from the road on the latter part of the way, but fresh and vigorous; for he was a powerfully built man, and was always careful to keep himself in good condition. He played football in those days, and boxed, and played tennis, and in other ways saw to it that his body as well as his mind got the exercise it needed. In winter he skated when there was ice,

but he had had few opportunities and was an indifferent per-
former. I remember an amusing little incident which shows
how thoroughly he was, as befitted his gifts, a man of theory.
He had a pair of straight-bladed skates, and he found that he
could not skate on the outside edge, as he wished to do. He
forthwith ordered a pair with curved blades, and demonstrated
that it was impossible to skate on the outside edge with the
straight blade. He was so very eager and so very sure, and the
demonstration was so far above my head, that he almost per-
suaded me, in spite of the fact that more than thirty years
before I had myself attained a modest degree of skill on the
ordinary straight blade. After all, we were in Birmingham :
what more natural than that I should be deluding myself and
others with a Brummagem imitation?

In those days we used to fill the house, especially at Easter,
with large and merry parties, friends of our children. They
were nearly all much younger than McLaren, but something of
the boy survived in him to the end, and he threw himself into
their amusements and played the part of elder brother to per-
fection. I am afraid he made himself quite useful in concealing
from me escapades which he—and others—thought it was
better for me, and for them, that I should not know. Whatever they wanted he would do his best to provide; he would
do anything to gratify a child's wish. One morning, looking
out of my bath-room window, I was astonished, and a trifle
anxious, to see him near the top of a tall beech. My youngest
daughter had told him that she wanted a rook's egg. As the
nest was there, he climbed for the egg—and got it. How he
overcame the difficulties of the big trunk I cannot tell, but I
do know that at one stage of the proceedings he was hanging
from a bough very high up by the arms alone. His arms, by the
way, were very powerful. I don't think he knew what fear was.
On one occasion we were cycling, Mrs Walker in front, he
behind her and I behind him. We were going pretty fast down
hill, when she fell. Before I well knew what had happened,
without checking speed he had somehow flung himself off his
cycle and with the cry, My God, was leaning over her. If he
could be absent-minded he could also be magnificently prompt.

In 1912 his father and mother and sisters visited England, and he was very happy in their company. The whole five visited Lampeter together. Bruce at first was not altogether willing that they should come. I think he was a little nervous as to how his Australian family might fit in with his English family. But the pre-established harmony proved to be an affair of families and not merely of individuals, and after the first day his nervousness completely disappeared. He became quite himself—absent-mindedness and all. I remember his English mother rebuking him before his Australian father for some omission which had occasioned confusion, and tying a thread round his finger so that he should not forget again. Old Mr McLaren was greatly amused. The glance, half laughing and half sheepish, which Bruce threw at me was that of the younger brother: he knew that I was often in the same scrape myself. A visit to Switzerland and Italy was part of the programme of the Australians, and the gallant old man—he must have been about 70 —devoted part of his time while he was with us to learning the Italian language, so that he might be better qualified to enjoy Italy. Such mental alertness in the father helps to explain the mental alertness of the son. Unfortunately, in Switzerland Mr McLaren became dangerously ill. Bruce hurried away after them, and most devotedly nursed his father back to health. He saw the whole family (except his brother in Korea) united once more when he visited Australia in 1914 along with the British Association; but his father died while Bruce was on the voyage back to England.

War had broken out before he landed, and he, like others, was faced with the fateful choice. He was very silent and very thoughtful. He did not ask advice—at least not of me. I think he knew that I thought and would have said that his own heart would give him the best advice. I am sure he never hesitated as to giving himself for service in some form. But a man so gifted might well hesitate as to the form. It may be that he, the accomplished mathematician, could have done more for the country outside the army than in it; but I imagine he felt that he could not rest away from the fighting line. He was no bantam, but fate decreed that he should be an officer in a bantam battalion. He was by this time Professor of Mathematics in Reading University College, and much of his military training was done there. I believe that his absent-mindedness sometimes showed itself amusingly—that he would in his abstraction cross over to the side-walk as he marched down the street with his platoon, leaving his men uncertain whether they should join their officer or keep the line he had just deserted. He visited us, at Birmingham, in the Christmas vacation of 1914–15, looking very brand-new in his uniform, and in June 1916 he spent part of his last leave with us at Lampeter. Like so many others, he showed little inclination to speak about the war, and we did not trouble him with questions about his own experiences. He loathed bloodshed and was altogether out of his element: only a stern sense of duty had taken him where he was. I believe he valued the deep peace of Lampeter during those few days— the very thing which had been almost intolerable to me the summer before, because of the poignant contrast with what I knew to be going on across the Channel. But he had earned the peace, while I had not ; only, I must confess with shame, in 1916 custom had blunted my sense of the contrast. He left us, and, after the manner of our kind, we were unemotional at parting. He was wounded on the 13th of August following while trying to save a dump of ammunition which had caught fire, and died on the 16th.

Some 800,000 men of these islands fell in the War. The bare number is appalling, but the number is not the worst. War spares the ripe and takes the green. Old men might have their hearts wrung with anguish by the very peace which they felt they had not earned, but they were not killed. It was "the expectancy and rose of the fair state" that fell. Of the 800,000, some 2500 belonged to McLaren's own University of Cambridge, and fully 600 to his College, Trinity. These were the flower of England, and among them McLaren was one of the most distinguished. Who shall say how much lies buried in that untimely grave ? There is no loss to humanity so heavy as the loss of leaders, and McLaren had the intellect to be a leader. Gifts devoted to such ends as his do not soon mature. He was just beginning to garner his sheaves, when war blighted the rich

promise of harvest. This volume, whatever be its value, is but
a fragment of what, we may be sure, he was about to achieve.

> He bore the load of thoughts that passed the spheres ;
>   Exile he bore, for duty must be done ;
> Few were his friends, and rarer still his peers;
>   Alone he stood, for genius lives alone.
>
> The world crashed round him ; and his soul, called back
>   From those "strange seas" whereon it voyaged still,
> Faced humble tasks to shape an Empire's track
>   One hair's breadth nearer the Eternal Will.
>
> He died. But sure that spirit pure and high
>   By death has made his own the immortal prize;
> For always, in the Everlasting's eye,
>   The grandest virtue is self-sacrifice.

OBITUARY NOTICE

By Prof. J. W. NICHOLSON, F.R.S.

*[Extracted from the Proceedings of the London Mathematical Society,
Series 2, Vol. 16, Part 5 (1916).]*

SAMUEL BRUCE M^cLAREN.

Samuel Bruce McLaren was born on August 16th, 1876, at
Yedo, in Japan. When he was about five years of age, his
parents, who were both Scotch, removed to Melbourne, where
his father, Samuel Gilfillan McLaren, had been appointed
Principal of a Presbyterian College. From his earliest years
McLaren showed great ability, carrying off the highest honours,
both at his school, the Scotch College, Melbourne, and at the
University of Melbourne. One of his teachers at the University stated in 1903 that McLaren was by far the ablest student
he had met during his twelve years' tenure of office, and one
whose ability should be sufficient to place him in a very conspicuous position as an original thinker. None who knew
McLaren intimately could doubt the truth of this verdict, for
few men gave so strong an impression of reserve power. Although there is evidence that the voluntary exile from his
domestic ties, which in his case were strong, cost him much, he
came to England in 1897, and entered Trinity College, Cambridge. He was elected into a Major Scholarship in 1899, being
Third Wrangler in the same year, and took his degree in 1900.
Taking Part 2 of the Mathematical Tripos in his third year,
at that time an unusual course, he was placed in the Second
Division of the First Class. In 1901 he was elected to an
Isaac Newton Studentship, and graduated M.A. in 1905, proceeding about this time to the University of Birmingham as
Lecturer in Mathematics*.

[* He was appointed Lecturer in Mathematics at University College, Bristol,
in 1904, leaving in 1906 to take up a similar appointment at the University of
Birmingham.]

At Cambridge he gave perhaps the impression rather of idleness than of industry, if judged by the quite fallacious standard of published papers. It was characteristic of McLaren that the subjects which absorbed his attention most completely were always those of a really fundamental nature. He never published a paper dealing with any problem whose importance was merely secondary, or in any sense only a matter of detail. The nature of the work which he produced at a later period leaves no doubt that the more fundamental problems of mathematical physics were taking shape in his mind during the Cambridge period, and had been pondered unceasingly in the light of the newer theories then springing up, which appeared to be undermining the whole foundation of the theory of radiation as ordinarily conceived. The opportunity of collecting his work into a connected scheme was presented to him when the subject of the Adams Prize Award for 1913 was announced. He was successful in this competition, and one of the tragedies of his untimely death is the fact that it occurred under circumstances which had precluded him from publishing his essay. It is not at present certain how much of it is in being, for he was occupied in its revision immediately before the outbreak of war. Much of his work may be lost irretrievably, and buried for ever in his grave at Abbéville. But even when our attention is confined, as perforce it must be, to the few papers which he actually published, his contribution towards the problem of radiation and of the nature of matter is found to be very considerable, and it is pervaded by a highly original point of view developed with great mathematical skill.

In 1913 McLaren was appointed to the Chair of Mathematics at University College, Reading, and, in the brief period before the outbreak of war, he threw himself completely into all the activities of a rising institution seeking an existence as an independent University. His personal popularity among his colleagues was very great, and many tributes have been paid by them to his memory. Previously somewhat aloof from the scientific societies whose scope included his own main interests, during this period he became more active in this regard. At the time of his death, he was a Secretary of Section A of the

British Association, and a peculiarly appropriate officer in view of the Australian visit. There is no doubt that he keenly appreciated this opportunity of returning to familiar scenes, marred, however, by the death of his father. The greatness of the shock which this was to him was known only to his more intimate friends.

This is not the appropriate place in which to give any full account of his work. His manuscripts are in the hands of those who have undertaken the task of saving as much as possible from oblivion, and a few general remarks are alone desirable, relating entirely to his published papers. These may be found in the *Philosophical Magazine* for 1911 and later years. In the first important paper, MᶜLaren discusses the emission and absorption of energy by electrons, and, in a very original manner, extends Lorentz' theory of complete radiation to all wave-lengths. The radiation is regarded as an external field superposed on the other forces which act on the electrons, and MᶜLaren points insistently to the essential fact that we must be at liberty to treat the radiation as a small disturbing force which gradually takes the electron out of the orbit it would follow under the internal forces alone. When this is ensured, we can use the theory of varied motion, the displacement due to radiation being small during the time of describing a free path, small in comparison with length of path and with any wave-length under consideration. MᶜLaren takes this as the key method in all his investigations, and develops it with considerable power. In this first paper, he points out that the exponential term in Wien's law belongs to the emission, and the paper concludes with a general formula, of the type of an integral equation, for the partition of energy in wave-length, to which any assigned distribution of electronic velocities can be fitted, and giving the Lorentz formula with the kinetic theory of distribution.

These ideas are developed further in a second paper, published a few months later. When the problem of energy distribution is thus reduced to the determination of a function f, it is found that the determination is beyond the reach of our present dynamical theory. If Hamilton's equations as derived from Least Action, and an essentially positive energy function, are used, we must have equipartition. We can give up Least Action and alter Hamilton's equations without coming nearer to a solution of the radiation problem, for no continuous laws of motion in which energy is conserved can account for the distribution of energy realised in nature, and the revolution foreshadowed by Planck and Einstein appears to become inevitable.

Such conclusions were, of course, reached by Jeans and Poincaré also, but McLaren's investigation proceeded on quite different lines. He did not like to assume that the statistical method is applicable to the æther with its infinity of degrees of freedom, and accordingly undertook, in a later paper, a direct calculation of emission and absorption, applicable to matter obeying any continuous laws of motion whatever. This paper is an attempt to save the classical view of radiation as a continuous wave motion. McLaren regarded it as a relatively small thing to sacrifice merely the ordinary mechanical notions of matter, being prepared to accept atomism for matter but not for radiation. In this paper he assumes the undulatory theory of light, but not the formulæ of classical dynamics. The laws of motion of matter need not now be continuous laws at all, though some at least of the variables specifying the conditions of matter must have a continuous range. Discontinuous changes must be governed by conditions not directly dependent on the presence of radiation, and, for example, may occur when velocities or coordinates reach prescribed values. Equipartition ensues as usual if the Principle of Least Action be assumed, but in the absence of this assumption, a formula very similar to Planck's can be derived on certain simple alternative hypotheses, although, as McLaren states, nothing is done to show that these hypotheses are actually consistent with possible laws of motion. The whole investigation is highly interesting and suggestive, and quite unlike anything to be found in other treatments of this question. McLaren at the same time shows that the classical dynamics cannot furnish an adequate kinetic theory of induced magnetisation.

His last published paper is in many respects the most striking, and begins by resuming, from a new standpoint, the discussion of a theory of gravity which goes back to Riemann, of æther as fluid, and of matter as a place of destruction gravity being an interference effect of the æthereal motions associated with portions of matter. The paper is very long, and an abstract cannot be attempted here. But McLaren passes on to the modern theory usually connected with Einstein, or rather to a theory independently derived, and of a similar nature. He was a believer in the physical reality of the fourth dimension, and went further than Minkowski. Though, as with Minkowski, the four-dimensional universe is without change, it is not, with McLaren, without time or motion. Time appears as a purely logical succession, the symbol of an order in which the elements of Minkowski's space are throughout.

McLaren presented some of the more fundamental parts of his work in this connection to the Mathematical Congress at Cambridge in 1912, and his results were in accordance with those of Abraham, but obtained independently by a quite different mode of procedure. He undoubtedly anticipated Einstein and Abraham in their suggestion of a variable velocity of light, with the consequent expressions for the energy and momentum of the gravitational field, and put forward, in his last paper, a suggestion, perhaps of great value, as to the nature of the magneton and of the elementary positive charge.

McLaren was singularly modest regarding his work, and this trait, combined with a great number of other interests, and a personality which was even boyish to the casual observer, tended to hide his real gifts from all but his more intimate friends and those familiar with his writings. On all the matters to which he gave greatest thought he was one of the most reserved of men. He had the highest sense of duty, and few know the mental crisis through which he went while in Australia with the British Association at the outbreak of war. He studied signalling on his way home, and having reached his decision, hesitated no further. We must admire such a decision in one for whom the military life had no attraction, but at the same time regret a national state of mind which could not suggest, to a man of this type, that the best national work he could do was in a different category. The parallel between his case and that of Moseley is very close, and they are perhaps the two most irreparable losses to the future of British science which this War has produced.

McLaren was, as an officer, idolised by his men, and was conspicuous for personal bravery in an army where bravery is almost universal. He met his death as the result of an endeavour to save a store of bombs threatened by fire. Though already badly wounded, and begged to desist, he went on and received the further wounds from which he died a few days later, on August 13th, 1916, at Abbéville. The bombs were not saved, and his own life, and with it much that future generations might have hoped for, was lost.

SECTION I

ACCOUNT OF PAPERS MAINLY ON RADIATION AND GRAVITATION

DRAWN UP BY PROF. J. W. NICHOLSON, F.R.S.

This section contains the substance of McLaren's Adams Prize Essay, largely represented, as it was, in his papers on Radiation, and arranged in the order of these papers, with the small departures from the text of those papers indicated as they arise.

The actual list of his papers on Radiation is as follows:

1. "Hamilton's equations and the Partition of Energy between Matter and Radiation." *Phil. Mag.*, Vol. 21, pp. 15–26 (1911).

2. "The Emission and Absorption of Energy by Electrons." *Phil. Mag.*, Vol. 22, pp. 66–83 (1911).

3. "The Emission and Absorption of Radiation in any Material System, and Complete Radiation." *Phil. Mag.*, Vol. 23, pp. 513–542 (1912).

4. "The Theory of Radiation." *Phil. Mag.*, Vol. 25, pp. 43–56 (1913).

5. "Aether, Matter, and Gravity." *Proc. Fifth International Congress of Mathematicians*, 1912, Vol. II, p. 264.

6. "A Theory of Gravity." *Phil. Mag.*, Vol. 26, pp. 636–673 (1913).

Of these papers, the third and fifth are not represented in the Essay, and are therefore not included in the following account. There are missing chapters for the Essay at the point at which this work of Paper 3 would naturally come, and it seems probable that these chapters were, originally, substantially identical with Paper 3, but that McLaren was dissatisfied and had destroyed them pending re-writing them in a new form. On such points we can only conjecture.

The absence of the fifth chapter is of less importance, as it is to some extent represented in his miscellaneous papers which are dealt with elsewhere in this volume.

HAMILTON'S EQUATIONS AND THE PARTITION OF ENERGY BETWEEN MATTER AND RADIATION

[*Phil. Mag.* (6), *Vol.* 21, *Jan.* 1911, *pp.* 15–26.]

This was McLaren's first important paper dealing with problems of radiation, and it is identical with a great part of the manuscript of his Adams Prize Essay. That manuscript, in this section, was apparently an older copy; for in the printed paper, the exposition has been improved in certain respects, and bears traces of later revision. But so far as the analysis is concerned, there is no real modification, and we shall accordingly give a brief digest of its nature, referring readers to the printed paper for a full development of a very difficult and notable piece of mathematical analysis.

The paper has a two-fold object. In the first place, Maxwell's theory of energy-partition, which McLaren evidently disliked always, is extended to forms of energy which are not quadratic in the velocities or momenta of the component parts of the system; and an attempt is made to bring within its scope the interaction of matter and radiation. He is content to suppose merely that if the energy is finite, so also are all the momenta, and quotes, as a characteristic instance, the energy of the Lorentz electron, when the momentum p occurs in the energy in the form

$$c \left(p^2 + m_0^2 c^2 \right)^{\frac{1}{2}},$$

where c is the velocity of light, and m_0 the mass for slow speeds.

After a discussion of the principles underlying the general dynamical method applied to a system whose initial momenta and values of the coordinates are mere matters of chance, with its result that in most cases the distribution of energy and momentum is near to the "temperature" distribution, he proceeds to discuss the partition of energy in a Hamiltonian system. In this preliminary statement, he lays down certain necessary postulates:

(1) The laws of heat are dynamical,

(2) Any set of values of the coordinates and momenta is possible if they are consistent with the constancy of energy,

(3) All such sets of values are equally probable.

In the general Hamiltonian system, the equations of motion are, as usual, typified by

$$\frac{dq_r}{dt} = \frac{\partial H}{\partial p_r}, \quad \frac{dp_r}{dt} = -\frac{\partial H}{\partial q_r};$$

but the expression for the kinetic energy requires definition when H is no longer quadratic in the momenta. McLaren defines the kinetic energy associated with a momentum p_r as

$$\tfrac{1}{2} p_r \frac{dH}{dp_r},$$

which is reducible to the usual form when H is quadratic. The Lagrangian function is

$$L = \Sigma p_r \frac{dH}{dp_r} - H,$$

so that

$$\tfrac{1}{2}(L + H) = \Sigma \tfrac{1}{2} p_r \frac{dH}{dp_r},$$

where, in the simple case, L is the difference of the kinetic and potential energies, and H is the sum. Thus the kinetic energy again appears here as the sum of terms defined as above. The " virial " of Clausius is

$$\Sigma \tfrac{1}{2} q_r \frac{dH}{dq_r}.$$

He now introduces the usual differential invariant

$$dN = dp_1 dp_2 \dots dp_n . dq_1 dq_2 \dots dq_n,$$

and imagines a complex of systems all satisfying the Hamiltonian equations but starting from every type of initial condition. As dN represents the number of systems having their coordinates and momenta between the limits (dp_r, dq_r) $(r = 1, \dots n)$, this distribution of number is invariant, and the group of systems which start with the same values of H will continue to possess these statistics. Each such group is on the average in a state of temperature equilibrium with the appropriate energy distribution. This distribution can be found by averaging all the systems.

The theorem of equipartition states that for all values of r and s,

$$\int p_r \frac{dH}{dp_r}\, dN = \int p_s \frac{dH}{dp_s}\, dN,$$

or, as he remarks, more simply

$$\iint p_r \frac{dH}{dp_r}\, dp_r dp_s = \iint p_s \frac{dH}{dp_s}\, dp_r dp_s,$$

and he proceeds to prove this from his specification.

McLaren also shows that

$$\int q_r \frac{dH}{dq_r}\, dN = \int p_s \frac{dH}{dp_s}\, dN,$$

from which the virial is seen to be equally distributed and equal to the kinetic energy. When H is quadratic, the potential and kinetic energies are equal and the virial becomes equal to the potential energy.

He also proves the four reciprocal properties

$$\int p_r \left(\frac{dH}{dp_s},\ \frac{dH}{dq_s}\right) dN = \int q_r \left(\frac{dH}{dp_s},\ \frac{dH}{dq_s}\right) = 0,$$

where r and s are different numbers.

McLaren now takes up the question, after these dynamical preliminaries, of the distribution of energy in matter and radiation, and a certain amount of definition of his notation is necessary. The æthereal variables he uses are the vector and scalar potentials $\mathbf{F}$ and ϕ. The vector velocity of electricity, of density ρ, is $\mathbf{u}$, and the equations of the æther are

$$\left(\nabla^2 - \frac{1}{c^2}\frac{d^2}{dt^2}\right)\phi + 4\pi\rho = 0,$$

$$\left(\nabla^2 - \frac{1}{c^2}\frac{d^2}{dt^2}\right)\mathbf{F} + 4\pi\rho\frac{\mathbf{u}}{c} = 0.$$

For any electron moving through it, let d/dt' denote differentiation at a point moving with it, and $\mathbf{E}'$ the whole electric intensity at its position. The electron is rigid, and if $\mathbf{p}$ is its true material momentum, its equation of motion is

$$\frac{d\mathbf{p}}{dt'} = \int \mathbf{E}'\rho\, dv,$$

where dv is an element of its volume. Also

$$\mathbf{E}' = -\frac{1}{c}\frac{d\mathbf{F}}{dt} - \nabla\phi - \frac{1}{c}[\mathbf{u},\, \mathrm{curl}\, \mathbf{F}]$$

in vector notation.

At the perfectly reflecting boundary, the tangential component of $\mathbf{E}$ is

$$-\frac{1}{c}\frac{d\mathbf{F}}{dt} - \nabla\phi,$$

and is zero, and everywhere

$$\mathrm{div}\,\mathbf{F} + \frac{1}{c}\frac{d\phi}{dt} = 0.$$

He remarks the indeterminateness of $\mathbf{F}$ and ϕ. If ω is any function, we can add $\nabla\omega$ to $\mathbf{F}$ and $-\dfrac{1}{c}\dfrac{d\omega}{dt}$ to ϕ, but a previous equation requires that

$$\left(\nabla^2 - \frac{1}{c^2}\frac{d^2}{dt^2}\right)\omega = 0.$$

Then ω can be chosen as a solution of this equation with any assigned value at the surface, and in particular so as to make ϕ vanish there. Thus the condition that $\mathbf{E}$ is normal to the surface implies also that $\mathbf{F}$ is normal.

This completes the specification of equations of motion for the æther and a typical contained electron. Much elegant analysis follows whose general scheme we shall alone indicate.

Let a function ψ be introduced such that

$$\mathbf{F} = \mathbf{F}_1 + \nabla\psi,$$

with $\qquad \mathrm{div}\,\mathbf{F}_1 = 0.$

Then

$$\mathrm{div}\,\mathbf{F} = \nabla^2\psi = -\frac{1}{c}\frac{d\phi}{dt},$$

and returning to the equations for the æther, we have

$$\nabla^2\left(\phi + \frac{1}{c}\frac{d\psi}{dt}\right) + 4\pi\rho = 0.$$

This shows that the expression in the bracket is the potential due to an electrostatic distribution of density ρ within the space.

If ψ is chosen to be zero at the surface, then since ϕ is zero also, $\nabla^2\psi$ is zero, and $\phi + \dfrac{1}{c}\dfrac{d\psi}{dt}$ is the potential due to a charge of density ρ and the charge it induces on the surface.

Since ψ is zero on the surface, the normality of $\mathbf{F}$ involves the normality of $\mathbf{F}_1$.

This surface condition, with

$$\mathbf{F} = \mathbf{F}_1 + \nabla\psi,$$

entitles us to expand $\mathbf{F}_1$ in a series of normal functions—a very beautiful feature of McLaren's mode of analysis. We write

$$\mathbf{F}_1 = \sum_{r=1}^{\infty} a_r \mathbf{F}_r,$$

where

$$\left.\begin{array}{c}
(\nabla^2 + k_r{}^2)\,\mathbf{F}_r = 0 \\[4pt]
\operatorname{div} \mathbf{F}_r = 0 \\[4pt]
\mathbf{F}_r \text{ is normal at the surface} \\[4pt]
\displaystyle\int F_r{}^2\,dV = 4\pi
\end{array}\right\}, \quad \ldots\ldots\ldots\ldots(A)$$

where dV is an element of the volume enclosed.

The functions $\mathbf{F}_r$ have the usual normal property

$$\int \mathbf{F}_r \mathbf{F}_s\,dV = 0, \quad (r \neq s)$$

(this is true even if an equality occurs between the numbers k_r, k_s). We easily find

$$\int \mathbf{F}_r \mathbf{F}_1\,dV = 4\pi a_r$$

and

$$\int \mathbf{F}_r \nabla^2 \mathbf{F}_1\,dV = \int \mathbf{F}_1 \nabla^2 \mathbf{F}_r\,dV + \int \left(\mathbf{F}_r \frac{d\mathbf{F}_1}{d\nu} - \mathbf{F}_1 \frac{d\mathbf{F}_r}{d\nu}\right) dS,$$

where S is the surface enclosing V, and the outward normal to S is used. This can be reduced easily to

$$\int \mathbf{F}_r \nabla^2 \mathbf{F}_1\,dV = \int \mathbf{F}_1 \nabla^2 \mathbf{F}_r\,dV = -4\pi a_r k_r{}^2 ;$$

and at the surface, we have the condition

$$\mathbf{F}_r \frac{d\mathbf{F}_1}{d\nu} = \mathbf{F}_1 \frac{d\mathbf{F}_r}{d\nu}$$

derived from (A) and the fact that $\operatorname{div} \mathbf{F}_1 = 0$.

We may now apply these results directly to the equation

$$\left(\nabla^2 - \frac{1}{c^2}\frac{d^2}{dt^2}\right)\mathbf{F} + 4\pi\rho\,\frac{\mathbf{u}}{c} = 0,$$

where $\mathbf{F} = \mathbf{F}_1 + \nabla\psi$. Thus

$$\left(\nabla^2 - \frac{1}{c^2}\frac{d^2}{dt^2}\right)\mathbf{F}_1 + \nabla\left(\nabla^2 - \frac{1}{c^2}\frac{d^2}{dt^2}\right)\psi + 4\pi\rho\,\frac{\mathbf{u}}{c} = 0.$$

Multiplying by $\mathbf{F}_r\,dV$ and integrating through the volume, using the equations just obtained,

$$-4\pi\left\{\frac{1}{c^2}\frac{d^2 a_r}{dt^2} + k_r^2 a_r\right\} + \int \mathbf{F}_r\nabla\left(\nabla^2 - \frac{1}{c^2}\frac{d^2}{dt^2}\right)\psi\,dV$$

$$+ 4\pi\int\rho\,\frac{\mathbf{u}}{c}\,\mathbf{F}_r\,dv = 0$$

(dv being an element of the volume containing electrons, whereas dV is an element of the whole volume).

Integration by parts shows that the term in ψ vanishes because of the surface conditions $\psi = 0$, $\nabla^2\psi = 0$, and therefore

$$\frac{1}{c^2}\frac{d^2 a_r}{dt^2} + k_r^2 a_r = \int\rho\,\frac{\mathbf{u}}{c}\,\mathbf{F}_r\,dv,$$

the integral being taken wherever charge exists. This is McLaren's first fundamental equation, determining the variation in time of the normal coefficients. The procedure by which it is derived is, in many ways, of great novelty, the whole analysis being much of a *tour de force*.

A further application of similar methods is then made to the determination of the equation of motion of any electron, from the basis of the formula we previously quoted for its rate of change of true momentum. For the full development, we must refer to the original paper, which is at this point identical with the manuscript. If dv_n is an element of volume of the nth electron—a typical one—McLaren writes

$$\frac{1}{c}\int\mathbf{F}_r\rho\,dv_n = \mathbf{F}_{rn},$$

where the integration is taken throughout the electron, and F_{rn} is thus a function only of the position of the nth electron.

If $\mathbf{u}_n$ is its vector velocity, our previous equation for the normal coefficients becomes

$$\frac{d}{dt}\left(\frac{1}{c^2}\frac{da_r}{dt}\right) + k_r{}^2 a_r = \sum_{n=1}^{n=N} \mathbf{u}_n F_{rn}, \quad \ldots\ldots\ldots\ldots\text{(B)}$$

and the new equation he derives is

$$\frac{d}{dt}\left\{\mathbf{p}_n + \sum_{r=1}^{\infty} a_r F_{rn}\right\} - \frac{d}{d\mathbf{q}_n}\left(\sum_{1}^{\infty} \mathbf{u}_n a_r F_{rn} - V_n\right) = 0, \ldots\text{(C)}$$

where $\mathbf{q}_n$ is the vector position of the electron.

It has been supposed that the electron is only in translational motion, each point having the vector velocity $\mathbf{u}_n$, and if

$$\phi_0 = \phi + \frac{1}{c}\frac{d\psi}{dt},$$

so that, as before,

$$\nabla^2 \phi_0 = -4\pi\rho,$$

and ϕ_0 can be regarded as a field of potential, the electrostatic force due to this field, on the electron, is

$$-\int \nabla\phi_0 \cdot \rho\, dv_n,$$

and V_n is the corresponding electrostatic energy of this field, dependent only on the position of the electron. (B) and (C) represent his final forms of the two fundamental equations.

There now arose, in McLaren's mind, the necessity for determining the relation of (B) and (C) to the general Hamiltonian dynamics, and some of his conclusions are of great interest. Let us suppose the momentum-vector $\mathbf{p}_n$ to be derived from a Lagrangian function L_0 for the *purely material* energy. Then the whole set of equations embodied in (B) and (C) has a Lagrangian function L, given by

$$L = L_0 - V_N + \sum_{r=1}^{r=\infty}\sum_{n=1}^{n=N} u_n a_r \mathbf{F}_{rn} + \sum_{r=1}^{r=\infty}\left\{\frac{1}{2c^2}\left(\frac{da_r}{dt}\right)^2 - \frac{1}{2}k_r{}^2 a_r{}^2\right\}.$$

We have remembered here that $\mathbf{F}_{rn}$ is a function only of q_n, and V_n is the whole electrostatic energy of the field of potential ϕ_0 already dealt with—the sum of the potential energies of all the electrons in the field ϕ_0.

The Hamiltonian function corresponding to this form is

$$H = \sum_{n=1}^{n=N} \mathbf{u}_n \frac{dL}{d\mathbf{u}_n} - L_0 + V_n + \sum_{r=1}^{r=\infty} \left\{ \frac{1}{2c^2} \left(\frac{da_r}{dt} \right)^2 + \frac{1}{2} k_r^2 a_r^2 \right\}.$$

Terms linear in u_n and a_r would be of a "gyrostatic" nature, but in fact they disappear.

Now let L_0 be zero. Then the momentum $\dfrac{dL}{du_n}$ does not contain the velocities. The system has $3N$ material momenta which depend only on a_r and q_n (as typical). Thus the velocities u_n cannot be expressed as functions of the corresponding momenta. Equation (C), in fact, can be made to show identical relations between the velocities, which precludes the use of the statistical method. Thus finally, L_0 cannot be zero if dynamical procedure is applicable to the system of electrons and æther.

When L_0 exists, on the other hand, Maxwell's method can be used. The equipartition then deduced is given by

$$\tfrac{1}{2} p_r \frac{dH}{dp_r} = \tfrac{1}{2} p_s \frac{dH}{dp_s},$$

where average values are used, or, with the same interpretation,

$$\tfrac{1}{2} u_r \frac{dL}{du_r} = \tfrac{1}{2} u_s \frac{dL}{du_s} \qquad (r \neq s).$$

If we write $\dfrac{da_r}{dt}$ for u_r and u_n for u_s, so that u_n is the component of $\mathbf{u}_n$ along the x axis,

$$\frac{1}{2c^2} \left(\frac{da_r}{dt} \right)^2 = \frac{1}{2} u_n \frac{dL_0}{du_n} + \frac{1}{2} u_n \sum_{r=1}^{r=\infty} a_r F_{rn},$$

or
$$\frac{1}{2c^2} \left(\frac{da_r}{dt} \right)^2 = \frac{1}{2} u_n \frac{dL_0}{du_n} + \frac{1}{2} \frac{u_n}{c} \int F_{1x} \rho \, dv_n,$$

where F_{1x} is the x component of F_1, and mean values are used. We are led to the conclusion that each degree of freedom of the æther has the same kinetic energy as each material degree of freedom.

McLaren anticipates the criticism that the real radiation is not what appears as radiation in this investigation, for the field just outside an electron should be regarded as part of its energy. But he points out that any attempt to discriminate between radiation and non-radiation leads us at once to a set of equations which a dynamical system cannot deal with or satisfy. He emphasises that he has only endeavoured to pursue the general dynamical method to its logical outcome for such systems.

THE EMISSION AND ABSORPTION OF ENERGY BY ELECTRONS

[*Phil. Mag.* (6), *Vol.* 22, *July* 1911, *pp.* 66–83.]

The fundamental object of this paper is the extension, to all wave-lengths, of Lorentz' theory of complete radiation. The electrons are regarded as moving in a conservative field of force due to positive charges, absorbing energy from the external radiation and emitting energy by virtue of their motion. Their velocities are not restricted to be small. Although not a believer in two aspects of Lorentz' work, in that he regards certain restrictions as unnecessary, McLaren adopts them for the time being, with the statement that these restrictions can be removed without altering the character of the results. Clearly he had at this time obtained a proof to this effect. The restrictions in question are (1) that the positive charges are assumed to be at rest, and (2) that the electrons are free from mutual influence.

The radiation is reduced, for simplicity, to a mere external field superposed on the other forces. A further subsidiary object of the paper is to indicate how the radiation may be differentiated from the forces of interaction of the positive or negative charges at any internal point of a hot body.

McLaren commences with a summary of the situation existing in the theory of complete radiation at that time, in the notation now familiar. If dN electrons have their coordinates (x, y, z) and momenta (p, q, r) within differential ranges $dx \ldots dp \ldots$, we write

$$dN = f(H)\, dp\, dq\, dr\, dx\, dy\, dz,$$

where $f(H)$ is a function of H, the energy of the electron at that point of the field. If $E_\lambda d\lambda$ is the complete radiation per unit volume between wave-lengths λ, $\lambda + d\lambda$, McLaren proposes to prove that

$$E_\lambda \int_{0_-}^{\infty} -\frac{d}{dH}\{\log f(H)\}\, \phi(\lambda)\, dH = \frac{8\pi}{\lambda^4} \int_0^{\infty} \phi(\lambda)\, dH,$$

where $\phi(\lambda)$ is a function of H and λ depending on the nature of the forces.

In the kinetic theory distribution, $f(H)$ is proportional to $e^{-H/R\theta}$ and the result reduces to

$$E_\lambda = 8\pi R\theta\lambda^{-4}$$

in agreement with Lorentz. E_λ is smaller if, for any reason, $f(H)$ decreases more rapidly as H increases. McLaren remarks that for small values of $\lambda\theta$, this is to be replaced by "some such" formula as Planck's

$$E_\lambda = 8\pi ch\lambda^{-5}\,(e^{ch/R\lambda\theta} - 1)^{-1},$$

but such a conclusion was certainly, at least then, unwelcome to him.

In spite of the success of the Planck formula, in whose deduction ordinary dynamical laws were discarded, McLaren registers a plea for the fundamental importance still left to the Lorentz method. For although dynamical laws could not lead to Planck's formula, it was not clear that they failed to allow the existence of the distribution it represented. Moreover, the method deals with processes known to be at work in absorption and radiation of metals, and which, for long waves, certainly bring about the distribution dynamics requires. The paper in fact shows that in so far as the electrons obey dynamics and the principles of kinetic theory, they must produce equipartition of the radiant energy.

At that time, somewhat loose conjectures were rife as to the modes in which the exponential factor in Planck's formula might arise, and the author removes them by a simple analytical instance which is typical. Radiation occurs if a force acts on an electron colliding with a positive charge. Such a force must, in a collision, reach a maximum and then decrease. Such a maximum occurs if the force is proportional to $(a^2 + t^2)^{-1}$. The radiation of period $2\pi/p$ is then a multiple of

$$\left(\int_{-\infty}^{\infty} e^{ipt}\,\frac{dt}{a^2 + t^2}\right) \times \left(\int_{-\infty}^{\infty} e^{-ipt}\,\frac{dt}{a^2 + t^2}\right),$$

containing the factor e^{-2pa}, and it is only necessary to make a a multiple of θ^{-1} to obtain the proper form. An application of McLaren's later an lysis, however, to this instance, shows that the factor comes also in the absorption, and thus disappears from the complete radiation formula. This result is the inevitable fate of any such suggestion.

We may now proceed to an account of the main features of McLaren's analysis in this paper, restricting our attention to his fundamental methods, and again referring back to the paper for the full investigation.

The author regards it as necessary that he should be free to treat the radiation "as a small disturbing force which gradually takes the electron out of the orbit it would follow under the internal forces alone. That ensured, we can use the theory of varied motion. The displacement due to radiation must be small during the time of describing a free path—small not only by comparison with the length of the path, but also a small fraction of any wave-length dealt with."

A free path, to McLaren, is the average distance moved by the electron before the direction of its velocity is completely altered. He gives an interesting discussion of the extent to which this deviation occurs under Planck's law of radiation. The matter is of some interest, in that we have seen no other discussion of a similar nature, and it is desirable therefore to reproduce it here in some detail.

If an electric intensity

$$\mathbf{a} \cos pt + \mathbf{b} \sin pt,$$

where $\mathbf{a}$ and $\mathbf{b}$ are vectorial, acts on an electron of mass m, and if $\mathbf{r}$ is the vector of position of the electron,

$$m\ddot{\mathbf{r}} = e\mathbf{a} \cos pt + e\mathbf{b} \sin pt,$$

and by means of two integrations,

$$\mathbf{r} - \mathbf{r}_0 - t\left(\frac{d\mathbf{r}}{dt}\right)_0 = -\frac{e\mathbf{a}}{mp^2}(\cos pt - 1) - \frac{e\mathbf{b}}{mp^2}(\sin pt - pt),$$

the suffix zero denoting an initial value, while the time t is for the free path. The expression on the left represents the deviation produced by the periodic force, or $\mathbf{r}_p$. If average values are denoted by a bracket $\{\ \}$,

$$\mathbf{r}_p{}^2 = \frac{e^2}{m^2 p^4}\{\mathbf{a}^2\}(\cos pt - 1)^2 + \frac{e^2}{m^2 p^4}\{\mathbf{b}^2\}(\sin pt - pt)^2.$$

The total electric and magnetic energy is, per unit volume,

$$\frac{1}{8\pi}\left[\{a^2\} + \{b^2\}\right].$$

If this is the energy in the interval between p, $p + dp$, then according to Planck's formula, since $\{a^2\} = \{b^2\}$, this average energy becomes

$$\frac{8\pi ch}{(2\pi c)^4}\, p^3\, (e^{hp/2\pi R\theta} - 1)^{-1}\, dp.$$

The average square of the total deviation in time t is the sum of such terms as $\{r_p^2\}$, and becomes

$$\frac{h}{c^3}\left(\frac{e}{m\pi}\right)^2 \int_0^\infty \frac{du}{u}(2 - 2\cos u - 2u\sin u + u^2)(e^{\alpha u} - 1)^{-1},$$

where $u = pt$, $\alpha = h\,(2\pi R\theta t)^{-1}$.

McLaren quotes Riecke to the effect that the length of the free path in metals is inversely proportional to $\theta^{\frac{1}{2}}$, while the average velocity of its description is proportional directly to $\theta^{\frac{1}{2}}$. It follows that θt is constant.

From the known values of the constants, and certain numerical data then available, such as the fact—based on the supposition that the collisions of the electrons are with single atoms—that for $\theta = 300°$, the free path is 10^{-6}, he finds that α is of order unity, the mean square of deviation is of order 10^{-23}, and the mean deviation for a free path is of order 10^{-11} at most.

McLaren then discusses the significance of Wien's law of radiation, from the standpoint that the time required in a free path is proportional inversely to the absolute temperature. Since Wien's law also gives the ratio of emission and absorption with the exponential factor $e^{-ch/R\theta\lambda}$ when $\lambda\theta$ is small, he raises the question as to whether this factor belongs to the emission or absorption. The experiments of Hagen and Rubens, as he shows, leave no doubt that it is the emission which contains the negative exponential factor—a factor which the calculation above shows to be of order unity when the period of the waves is the same as the time spent in its free path by an electron. The rapid decrease of emission for smaller periods leads inevitably to a formula like Planck's and away from that of Lorentz.

McLaren was, of course, in conclusions of this nature, not putting them forward for the first time. Jeans, for instance, had done the same. But his point of view is essentially, and in fact, extremely individual, and so different, in its procedure, from that of others, that it may well be regarded as a fundamental contribution which at that time was much needed, and which did throw much new light on a very perplexing problem, and on the necessity for conclusions so very unwelcome to the majority of physicists at that time.

He now points out the essentially artificial character of the ideas of free paths and collisions of electrons moving in a metal. An electron is not to be imagined as completely deflected when it meets an atom, and moving uniformly till it meets the next, for the character of the radiation to which such a conception leads is quite at variance with the facts. For the bulk of the energy emitted would be, of necessity, of a period equal to the time of a collision, or about 10^{-15} seconds. Up to waves of this period, moreover, the emission would increase with decreasing wave-length. The contrast between free path and collision must, in fact, be avoided. So he now proposes a new definition of free path—as the average distance an electron must move before its velocity retains "no observable" connection with its original value. The main emission is then of periods not small in comparison with the time of a free path—Fourier analysis alone, without dynamics, predicts this. Thus Wien's law takes on an interesting aspect—it makes the maximum radiation occur at a wave-length λ_m inversely proportional to θ, because of the experimental fact that the time for a free path varies in this way. But it is of no service to construct laws of force which make the emission a function of θ, for all dynamical principles lead inevitably to the conclusion that the absorption and emission are of the same type, and the final conclusion is always the "unwelcome paradox" of equipartition.

After this introductory discussion, McLaren proceeds to the complete analytical treatment of the emission and absorption. The analysis is elegant and of great power, but to abstract it, even in comparatively small detail, would not be possible in an account of the nature here contemplated. The exposition in his

paper could not be improved upon, and he was clearly satisfied
with it also. It ends with the formula we quoted at the beginning of this account.

THE THEORY OF RADIATION

[*Phil. Mag.* (6), *Vol.* 25, *Jan.* 1913, *pp.* 43–56.]

In this paper McLaren makes a final attempt to save the
classical theory of radiation as a continuous wave motion. He
considered that if this could be done, the sacrifice of our
mechanical notions of matter would be a small thing, for even
its most elementary properties, such as the existence in chemical
elements of inviolable forms separated by gaps, seemed to him
to need essentially a concept of the general type of Einstein's
quantum. He declares himself prepared, at this point, to accept
atomism for matter, but not for radiation, which still remains
necessarily a continuous wave motion.

The method he now proposes to adopt is to discard the
formulæ of ordinary dynamics—or more precisely, to refuse to
assume them—but to retain the ordinary undulatory theory
of light. His method is thus at the same time simpler and
more general in scope than that of the first two papers. He
finds it possible to give a mathematical theory of emission,
absorption, complete radiation, and in fact of the whole process
of interaction between radiation and material systems, without
supposing that the laws of motion of matter are to be derived
from the Principle of Least Action, or even that they are
continuous laws of motion at all. Only certain of the variables
specifying the state of matter must be capable of a continuous
range, in order to fit in with the conservation of energy and the
continuous theory of radiation.

The material system he discusses is exposed to radiation in
an enclosure completely bounded with reflecting walls. The
normal coordinates $(a_1 \dots a_n)$, defined before, are again used, and
are now infinite in number. For a wave-length $2\pi/k_m$, the
differential equation

$$\frac{1}{c^2}\frac{d^2 a_m}{dt^2} + k_m^2 a_m = \phi_m$$

is presumed, where $m = 1, 2, \ldots \infty$, and ϕ_m is a function of the material state at time t. The field of radiation is determined by the values of all the a's. We can make a_m simple harmonic for longer and longer times by increasing the enclosed space. When it is so from time $t = t_0$ to time $t = t$, if zero suffix denotes the value of t_0, then when $t - t_0$ is small enough,

$$a_n = (a_n)_0 \cos (ck_m t - ck_m t_0) + \left(\frac{da_m}{dt}\right)_0 \sin (ck_m t - ck_m t_0).$$

The previous paper showed how to find a_m and ϕ_m on the basis of the electromagnetic theory of Lorentz. In our abstract of the paper, we gave, as an illustration, the analysis for the determination of a_m.

Possible discontinuous changes in the material system are dealt with as follows:

"The variables defining the state of the material system I denote by $x_1, x_2, \ldots x_n$. These have all a continuous though not necessarily an unrestricted range of variation. If discontinuous changes are admitted, we must suppose them to involve a transformation in the *nature* of these variables. The values of the new and the old must be connected by equations sufficient to determine the one when the others are given."

The italics in this statement are our own. It is clear that although this is certainly the most natural way to introduce discontinuity into an investigation of this nature, it is not the only possible way.

The possibility of a material state between the limits $dx_1 \ldots dx_n$ is

$$dN = \rho' dx_1 dx_2 \ldots dx_n = \rho' dV$$

for brevity, where ρ' is a certain function. Whether the change be continuous or not, if a material system changes from the time t_1 to the time t_2, we have

$$(dN)_2 = (dN)_1,$$

and the most general solution of these two equations is

$$\rho' = e^{-hH} \rho,$$

where ρ is any particular solution, and h is a constant. The greater generality of the laws contemplated does not in fact alter the law given by the statistical method. H is again, of course, the total energy, and we may assume that any quantity which remains invariant during the changes must be a function of H.

Since ρ' and ρ are both solutions,

$$(\rho'/\rho)_2 = (\rho'/\rho)_1,$$

and ρ'/ρ being thus an invariant, we have

$$\rho' = \rho f(H).$$

The statistical method lays down essentially that for thermodynamic purposes, a finite body can be divided into finite physically independent parts. Let there be two parts A and B, and then

$$dN = (dN)_A (dN)_B.$$

But $$(dN)_A = \rho_A f(H_A)(dV)_A,$$

and so for $(dN)_B$.

Thus

$$dN = f(H_A) f(H_B) \rho_A \rho_B dV_A dV_B,$$

and its invariance, involving the invariance of $\rho_A dV_A$, $\rho_B dV_B$, gives at once

$$(\rho_A \rho_B dV_A dV_B)_2 = (\rho_A \rho_B dV_A dV_B)_1,$$

so that this function is a special value of dN. But the general solution

$$dN = f(H_A) f(H_B) \rho_A \rho_B dV_A dV_B$$

then leads at once to the fact that $f(H_A) f(H_B)$ is a function of the total energy, or of $H_A + H_B$. Accordingly

$$f(H) \propto e^{-hH}.$$

Clearly the possibility of discontinuity does not enter, and McLaren's investigation makes a fundamental start. The preliminary statement regarding his views at this time of the interaction of matter and radiation must now receive our attention.

The distribution just discussed is no longer strictly valid in the presence of radiation, which also disturbs the motion of the material system. McLaren again proposes, as before, to discuss the radiation as a small disturbing agency acting on the normal system, producing first order variations in its characteristics.

The normal system he describes as "undisturbed motion," for convenience, and the new coordinates x_r are of the form $x_r + \delta x_r$, where $r = 1, 2, \ldots$ and δx_r is due to the radiation. The values of these increments are calculated directly by allowing the material systems distributed according to the statistical law e^{-hH} to move long enough in the presence of radiation to obtain its full effect. This duration of time defines the free path, and the deviation which occurs during this short interval is alone correlated with the disturbing forces at any instant. If ϕ is any function of the x's, and $\delta\phi$ its increment due to their increments $\delta x_1, \delta x_2, \ldots$, we have

$$\delta\phi = \sum_{r=1} \frac{\partial\phi}{\partial x_r} \delta x_r,$$

where δx_r, etc. are functions of $(x_1)_0$, etc. and of the time which appears explicitly in the general expression of the second equation we quoted.

The average value of $\delta\phi$ is readily found during a free path. For if we start at $t = t_0$ with a distribution such that the frequency dN_0 for the volume element dV_0 is

$$dN_0 = \rho_0 e^{-hH_0} dV_0,$$

and let radiation act till time $t = t$, the average value of $\delta\phi$ at this time is

$$\int^N \delta\phi \, dN_0.$$

But in the undisturbed motion, $dN_0 = dN$, and the average value of $\delta\phi$ is also

$$\int^N \delta\phi \, dN,$$

where the integrations in dN are over the range of the variables $x_1, x_2, \ldots$ in the undisturbed motion.

Now

$$\int \delta\phi \, dN = \delta \int \phi \, dN - \int \phi \delta \, (dN).$$

This fact is independent of any considerations regarding the continuity of the motion. If the radiation does in fact precipitate a sudden transformation of the variables, $\delta(dN)$ involves a finite change in $(x_1, x_2, \ldots)$ but only an infinitesimal one in dN.

In ordinary mechanics

$$\delta \int \phi \, dN = 0,$$

for here the variables are coordinates of position and momentum, and radiation does not affect their range of values. The limits of integration are unaltered, so this equation is true. McLaren uses this equation throughout, thereby, as he says: "It is thus assumed that if there are discontinuous changes, they are governed by conditions not directly dependent on the presence of radiation. For example, they may take place when the velocities or coordinates reach certain given values."

The formula for the average of $\delta\phi$ now becomes

$$\int \delta\phi \, dN = - \int \phi \delta \, (dN),$$

and it should at this point be clear just how McLaren is able to take account of the possibility of these discontinuous or catastrophic changes in an orderly analysis.

Interesting questions are raised when $\delta (dN) = 0$, for in these circumstances, the average value of $\delta\phi$ is zero, and the radiation produces no effects at all capable of being observed. The striking case occurs when the disturbing force is a steady magnetic field, as probably Voigt was the first to perceive. McLaren deduces at once that there can never be a kinetic theory of induced magnetisation, and clearly perceives that this fact in itself is a very severe blow to the classical dynamics. He returns to the subject at the end of the present paper.

In general,

$$\delta (dN) = (dN)_t - (dN)_0,$$

but $dN_0 = dN$, so that $\delta (dN)$ is in fact the difference in value of dN due to the disturbance, or

$$\delta (dN) = \int_{t_0}^{t} \frac{d}{dt} (dN) \, dt,$$

where the integrand is zero in the undisturbed motion.

McLaren proceeds to show that $\frac{d}{dt} (dN)$ is necessarily linear in a_m, da_m/dt, in order to obtain an absorption proportional, as it is, to the intensity of the radiation. Its full value is also dependent on the x's.

The full basis of the analysis has now been described, with a view to giving a comprehensive view of its nature. The remainder of the paper consists of the detailed working out, which is purely a matter of mathematical analysis of some complexity, but quite unsuitable for abstraction. We shall therefore pass directly to the ultimate result which McLaren obtains for the complete radiation as the rates of emission and absorption, separately calculated. This result is

$$E_\lambda = 8\pi R\theta\lambda^{-4}\,\frac{\phi_m}{\phi_m + \psi_m},$$

where (ϕ_m, ψ_m) are certain functions. All that the analysis, when equipartition is excluded, proves about them without any further assumptions not specified above, is that they have the forms

$$\phi_m = \int_0^\infty f(\lambda, H)\, e^{-H/R\theta}\, dH,$$

$$\psi_m = \int_0^\infty g(\lambda, H)\, e^{-H/R\theta}\, dH,$$

f and g being unknown. The essential difference, in an analytical sense, between the results of the Principle of Least Action and those obtained by discarding it, is that the function ρ—at the beginning of this account—is equal to unity in the former case.

The form of E_λ suggests, if vaguely, a possible fit with Planck's formula under special further assumptions, as McLaren remarks. Ordinary mechanical theories which explain radiation as due to the accelerated motion of electrons always contain, in the function which $f(\lambda H)$ actually is, some factor like $e^{-a/\lambda H}$, where a is constant, though this function $f(\lambda, H)$ is in such cases a very special form of McLaren's, in which λ and H do not necessarily occur as a product. However, to discuss further the possibility of a correspondence, he decides to take

$$f(\lambda, H) = f(\lambda H), \quad g(\lambda, H) = \lambda F(\lambda H)$$

for both functions, and further to suppose that they behave in a similar manner to the above exponential, i.e. with an essential singularity and vanishing with H. He further supposes a series exists of the form

$$F(\lambda H) = \sum_{n=0}^\infty a_n \frac{d^n}{d(\lambda H)^n} f(\lambda H).$$

Then $\qquad \psi_m = \lambda\theta \displaystyle\int_0^\infty F(\lambda H) e^{-H/R\theta} dH$

$$= \lambda\theta \sum_0^\infty (-)^n \frac{a_n}{(R\lambda\theta)^n} \int_0^\infty f(\lambda H) e^{-H/R\theta} dH,$$

and $\qquad E_\lambda = 8\pi R\theta\lambda^{-4} \left\{ 1 + \lambda\theta \sum_0^\infty (-)^n \frac{a_n}{(R\lambda\theta)^n} \right\}^{-1}.$

If we make the special assumptions

$$a_0 = 0, \quad a_n = \frac{k_1}{n!} (-kR)^n,$$

we get $\qquad E_\lambda = 8\pi R\theta\lambda^{-4} \{ 1 + k_1\lambda\theta (e^{k/\lambda\theta} - 1) \}^{-1},$

which is similar to Planck's formula. Such special artifices show at least that a wave theory of radiation, without the use of the Principle of Least Action, is a formal possibility. But it is nothing more, for as McLaren says, nothing has been done in the vital matter of showing even that these later assumptions are consistent with any continuous laws of motion whatever. And at this point he leaves the subject.

There are two final sections of this paper, only briefly indicated and the further development postponed, but in fact never given and almost certainly lost.

The first, which is described as the theory of a simple wave train, is intended to show that we can calculate the effect of radiation on matter when the laws of motion are widely different from those of dynamics. For with a vector potential $\mathbf{F}_1$ satisfying

$$\left(\nabla^2 - \frac{1}{c^2}\frac{d^2}{dt^2} \right) \mathbf{F}_1 + 4\pi\rho\frac{\mathbf{u}}{c} = 0,$$

which excludes all irregular interatomic fields—the proof of this is in an earlier paper—we have in an advancing wave

$$\mathbf{F}_1 = (f,\, g,\, 0)\, e^{ipt - inx - kz},$$

and thus $\qquad \left(\dfrac{p^2}{c^2} + k^2 - n^2 + 2ink \right) \mathbf{F}_1 + 4\pi\rho\dfrac{\mathbf{u}}{c} = 0.$

If we take dS as an element of area parallel to the wave front, dz a line element perpendicular to it, small compared with the wave-length, then restricting ourselves to the disturbance due to the wave

$$\left(\frac{p^2}{c^2} + k^2 - n^2 + 2ink\right) \mathbf{F}_1 dS\, dz + \frac{4\pi}{c}\, \delta \int \rho \mathbf{u}\, dS\, dz = 0,$$

and the average value of $\delta \int \rho \mathbf{u}\, dS\, dz$ is proportional to $dS\, dz$. The methods of the whole paper can then be applied.

The final small section goes further into detail, with an elegant analytical investigation, on the impossibility of explaining magnetic phenomena by the principles of classical dynamics. It is in fact an analysis, in his notation, of the effect of a steady magnetic field, but its procedure is not, in essentials, different from that of others.

A THEORY OF GRAVITY

[*Phil. Mag.* (6), *Vol.* 26, *Oct.* 1913, *pp.* 636–673.]

In this paper—the last paper of any length written by McLaren—perhaps the most original part of his general point of view is set forth. It follows his manuscript Adams Prize Essay very closely, and gives the best view of his trend of thought at the time at which we must take leave of it. He begins by directing our attention to Riemann, and his view that "æther" is merely a "fluid" and "matter" a region where fluid is destroyed. To the consideration of this view, he brings Einstein's principle, then new.

On this view, of course, matter is a sink—in the hydrodynamical sense—of æther, which flows in with radial symmetry to replace the destroyed æther. Two finite portions of matter disturb the flow of æther due to each other, and gravity, following Le Sage, is an interference phenomenon.

McLaren wishes to suppose that matter may be postulated as freely transferable through an omnipresent æther which cannot press upon it as upon a foreign body. When Newton's apple falls, it is carried down by a stream of æther, but does not fall only on this account. The momentum of the apple as it falls is derived from the destroyed æther, but not by virtue of a mechanical pressure due to the æther. If matter is an æther-source, and not a sink, Newton's law must hold. For the apple is in an ascending æther-stream which the earth creates. It must be driven down by the reaction against its own upward jet of æther.

McLaren considers that, if Riemann's ideas gain credit for a plausible and real simplicity, in their interpretation of the Newtonian law, only the drastic method of the relativist will solve their outstanding problems. Even if gravity is merely mass-motion of æther, what is to be said of optics? For an æther, capable of polarisation, must have some properties which he calls "quasi-elastic," in the hope that these properties may lead to light-waves. He states at once the fundamental problem for Riemann—why is the path of a ray of light and its velocity of propagation quite independent of all motion of the medium through which transmission takes place?

He attempts to deal with it as follows:

Take a function J which, with its first differential coefficients, is everywhere finite and continuous. It is a solution of

$$\left(\nabla^2 - \frac{1}{c^2}\frac{\partial^2}{\partial t^2}\right)J + 4\pi m = 0,$$

where m is an absolute constant (or weight) and does not depend on the type of matter. An element dv of matter experiences a force

$$m\,dv\,\nabla J = m\,dv\,N,$$

if N is the gravitational force of Newton. Let $\mathbf{u}$ be the (vectorial) velocity of æther of density ρ. If the motion is irrotational, in the sense of relativity theory,

$$\rho\mathbf{u} = \nabla J, \quad \rho c = -\frac{\partial J}{c\partial t}.$$

It is clear that J is a velocity potential, in the ordinary hydrodynamical sense. We can derive

$$\frac{\partial \rho}{\partial t} + \operatorname{div}(\rho\mathbf{u}) = -4\pi m,$$

which, of course, is the equation of continuity for the æther as fluid. The quantity $4\pi m$ of æther disappears per second for unit volume occupied by matter.

3—2

McLaren clearly does not wish to bind himself to any one of the possible constitutions of matter open to him at this point. He desires to treat matter as a manifestation of one "ultimate" substance, which is a fluid whose density and motion are at all points continuous. It has two forms, "matter" and "æther"—free "æther," in later terminology—which are mutually exclusive at any point of space.

If the fluid increases or decreases, we have matter, and it involves increase or decrease of momentum, derived from the free æther around, but belonging, like all momentum, to the ultimate pervading fluid. If this fluid decays or grows, an external source of energy is implied, and the conservation of energy—regarded as a mathematical fiction with no precisely-expressible physical meaning—compels us to attach to matter an energy per unit volume of magnitude $-mJ$.

In "æther"—free æther, as now called—growth or decay is impossible, but the medium can have polar properties. Maxwell's equations, in the usual notation,

$$\frac{dE}{dt} = c \operatorname{curl} H, \quad \frac{dH}{dt} = -c \operatorname{curl} E,$$

$$\operatorname{div} E = 0, \quad \operatorname{div} H = 0,$$

where (E, H) are electric and magnetic forces, as vectors, define such properties.

McLaren next proceeds to formulate boundary conditions at the confines of matter and æther. The ultimate fluid is presumed to pass freely from one form to the other, in *continuous* motion. The vectors E and H are automatically presumed, in consequence, to have the relations appropriate to a perfectly reflecting surface of separation. Moreover,

$$\frac{1}{8\pi}(E^2 - H^2) + mJ_0 = a - V_a,$$

where the conservation of energy and momentum presupposes this equation. A is a constant, V_a a certain potential function which, in a purely mathematical theory, is zero. (See § 10 of the paper.) The function J_0 is not quite J itself, the velocity potential, but differs from it by a function proportional to the

time. For the origin of J_0 we must refer the reader to the
original paper. In fact,

$$J = J_0 - \rho_0 c^2 t,$$

if t is the time, and ρ_0 is the density of the æther at infinity.
The true Newtonian potential is J_0, which is formally shown
to have this difference from J, the velocity potential of the
æthereal motion.

Now the motion of the fluid was defined by

$$\rho \mathbf{u} = \nabla J, \quad \rho c = - \frac{\partial J}{c \partial t}.$$

What can be the dynamical nature of the universal fluid? It
would fit the minimum action principle

$$\delta \iiiint L\, dv\, dt = 0, \quad L = \tfrac{1}{2} \rho^2 (\mathbf{u}^2 - c^2)$$

—on relativistic hydrodynamics—in the absence of circulation.
It can therefore exist as a motion in an ideal fluid. It remains
to be seen that the passage of a wave of light, and the presence,
in consequence, of electric and magnetic stresses, do not alter
the type of motion. McLaren raises a deeper question at this
point—does the motion itself change the form of Maxwell's
equations, since not only the velocity of the æther is a function
of the space-coordinates, but even its density?

He states that he meets these problems as a believer in the
physical reality of a fourth dimension—a convert to the view
that Minkowski did not raise a fiction which, in the mathematical sense, was effective after the manner that ordinary
electric images in electrostatics are effective, but raised instead
a structure which, *as reality*, is a necessary inference from ordinary experience. Few physicists, and perhaps, least of all,
ourselves, would agree, but McLaren's development of his ideas
is subtle and, taking the part of an advocate, we shall continue
his exposition.

In Minkowski's view, the four-dimensional universe is devoid
of change. In McLaren's, it is not devoid of time or motion.
Time, however, is a logical succession or order, and nothing
more—the symbol of an order in which the elements of

Minkowski's space are throughout, as McLaren puts it briefly. The symbol $d\tau$ is used by McLaren for an element of time, which he is careful to define as the symbol of an infinitesimal point-transformation.

Let us take the æthereal fluid as incompressible. Then the universe is defined by the minimum action principle

$$\delta \iiiint L\, dv\, ds\, d\tau = 0,$$

$$L = \frac{1}{2}\left(\frac{d\mathbf{r}}{d\tau}\right)^2 - \frac{1}{2}\left(\frac{ds}{d\tau}\right)^2 + \frac{1}{8\tau}(H^2 - E^2) + V,$$

$\mathbf{r}$ is a vector and s a scalar, and they give four variables. The element $dv\, ds$ is of volume, invariable in free æther, exponentially increasing or decreasing in matter. McLaren gives no reason for this supposition of an exponential law, but it is a reasonable expectation. We have

$$\delta\{dv\, ds\, e^{\sqrt{4\pi \cdot m\tau}}\} = 0.$$

The function V is the potential of any external field. The electromagnetic energy involved in the function L is quite statical. In free æther, it gives the usual force on any unit volume-element of magnitude

$$\nabla (H^2 - E^2)/8\pi.$$

This would not make the irrotational motion change its type, but would affect the distribution of pressure. It is an addition to the Maxwell stress-system, which he shows to give zero pressure and stress across the surface of matter. This force is thus, in a sense, irrelevant.

McLaren, in an argument to which we cannot make detailed reference, deals with the question of local aether-time, and finds it proportional to the velocity potential J. In fact

$$t_a = - J/c^2,$$

and there is a local æther density

$$\rho_a = \rho \sqrt{1 - u^2/c^2}.$$

McLaren had continued, in this direction, much further—in a manner in which his manuscript leaves no real trace. But he had worked out the energy and momentum which follow from these specifications and anticipated, at the 1912 Mathematical Congress, the results of Abraham. If we go past the matter suppressed from his manuscripts and from this paper, we find that he arrives at the conclusion that the velocity of light at any point of space is inversely as the density—(ρ_a in the last formula)—of the æther. Here he anticipates the suggestions of Einstein and Abraham, and others, which preceded the general Einstein principle. It is very unfortunate that his analysis should be, apparently, permanently lost to a great extent.

McLaren's use of time differs much from Minkowski's, for with him it is an independent variable described as the "absolute time" of an unchanging universe—the symbol, as we said before, of a continuous point-transformation, where any transformation generates a continuous succession of instants. He follows Robb in his discussion of "before" and "after." If t_p, t_a are the times at P and Q, whose vector coordinates are $\mathbf{r}_P$, $\mathbf{r}_Q$, and if

$$(r_P - r_Q)^2 - c^2(t_P - t_Q)^2 > 0,$$

then P is neither before nor after Q, while if

$$(r_P - r_Q)^2 - c^2(t_P - t_Q)^2 < 0,$$

then P is before or after Q according as t_Q is greater or less than t_P. These expressions are invariant under the Lorentz-Einstein substitution, and the relations of "before" and "after" are correspondingly invariant.

McLaren proceeds to develop a four-dimensional electro-dynamics, for which reference must be made directly to his paper. The origin of gravitation, in this work, appears as the momentum of destroyed æther, which is handed over to matter, and he has to consider how this process can be carried on without any interference with the irrotational flow of æther. Each element of æther, in its destruction, has to drop completely out, with corresponding loss of its momentum.

Returning finally to the question of moving media, McLaren in this paper seeks further to define his position. Essentially it stands upon the physical reality of a four-dimensional universe, and upon the inadequacy of any attempt to develop the theory in terms of the three-dimensional which our modes of thought accustom us to. He considers that his last paper has reached the limits of possible mechanical theory in regard to gravitation. The account we have given of his work has perhaps also reached the limit of what is possible in somewhat general terms, for no account can be really effective without a considerable body of very sustained mathematical analysis. We have given such indications of the great power of M°Laren, as an analyst, as should serve to direct the attention of those whose interests are mathematical, to a very orderly scheme of ideas in his mind which foreshadowed much that has happened since. He was unquestionably a real pioneer who worked on his own lines and borrowed little from contemporary work in his pursuit of a very individualistic standpoint. Though much of his essay is lost and cannot be replaced, enough remains for a fairly clear picture.

SECTION II. ELECTROMAGNETIC THEORY

[EDITORIAL NOTE: BY PROF. H. R. HASSÉ.

The papers in this section deal with a particular line of investigation which has its starting point in the paper "A Theory of Gravity" referred to in Section I above. In considering in this paper the existence of closed surfaces as the boundaries of matter, McLaren suggests a physical interpretation of the difference between simply-connected and multiply-connected surfaces. He provisionally assigns a negative sign to the electric charge on a simply connected surface, and a positive sign to that on a multiply-connected surface, or magneton. He shows that in the former the electric induction over the surface is constant, and that in the latter the magnetic induction through any aperture of the multiply-connected surface is constant.

These ideas were developed further in three short notes published in the autumn of 1913, and reprinted here as they indicate the lines on which McLaren was working at that time.

Unfortunately the theory based on these hypotheses was not completed before the War broke out, and it is not possible from the manuscripts left by McLaren, which are often merely preliminary in character, to say how he intended to carry out the further development of the theory. There is very little indication as to the ultimate scope of the investigation, or of the physical ideas involved, though it is clear from his letter to *Nature* (see below) that McLaren had in hand an ambitious programme, nothing less in fact than an electromagnetic vortex theory of matter.

The manuscripts may be divided into two parts, one, which is fairly complete, dealing with magnetic theory, and the other with the mathematical theory of the magneton. The substance of these researches, so far as that is possible, is given in this section, and an examination of them will emphasise the loss, already realised by all who are acquainted with his published papers, which Physical Science has suffered from the early death of their author.]

A THEORY OF MAGNETS

[*Transactions of Section A. British Association Meeting at Birmingham*, 1913.]

It is my object to recall some difficulties of magnetic theory and to suggest how they may be escaped. The history of magnetic science divides into an ancient and a modern period, the times before and after Ampère. In the earlier period the fundamental idea used is that of a magnetic substance; after Ampère this idea disappears. The magnet is now regarded as a whirl of electric particles or electric fluid. In modern electromagnetic theory all substance is electric; with Lorentz, for example, matter is the electric fluid.

Thus Ampère may claim to have given what previously did not exist, a theory of magnets. Before him the existence of molecular magnets was the starting point; any explanation of magnetic phenomena, as, for example, Poisson's account of magnetic induction, has to begin with matter whose elements are already magnetised. Poisson, I may remind the reader, can only account for paramagnetism; of diamagnetic phenomena there is no obvious explanation. I wish to point out that the modern electromagnetic theory has its own difficulties. It cannot take over unmodified Poisson's way of viewing the phenomena of magnetic induction.

It is true that the magnetic field due to a magnet is the same as that due to a rotating electric charge; it is true also that the resultant force exerted by the field on the charge is the same as that which it exerts upon the magnet. But it by no means follows that this force produces the same results in the two cases. And one fact sufficiently establishes a difference. A magnetic field does work upon the magnet in changing the direction of its axis, or the magnet in that field has potential energy; on the other hand, the same field can do no work by changing the axis of rotation of the electric charge, because the force on each element is at right angles to its motion. Take in particular a spherical distribution of charge rotating about any diameter. Set up a steady magnetic field, the only effect is to superimpose upon the original rotation about an axis fixed in the charge another rotation about the magnetic lines of force, with an angular velocity simply proportional to that force. This second motion accounts for diamagnetism, but there is no tendency for the axis of rotation fixed in the charge to approach the magnetic lines. The diamagnetic field is independent, as it ought to be, of the temperature, but no paramagnetic field is created.

The difficulty has been remarked by Lorentz, Voigt, J. J. Thomson, N. Bohr, and, I doubt not, by others. It is not, however, generally recognised that in such a theory of magnetic induction as Langevin's we are back at the ancient postulate of magnetic substance.

The difficulty may, I think, be evaded by moving still further from the old point of view as well as by returning to it. We may give up not only magnetic substance, but electric substance as well.

I assume the electromagnetic field defined by two vectors $\mathbf{E}$ and $\mathbf{H}$. This field is not all space; it is bounded by closed surfaces within which $\mathbf{E}$ and $\mathbf{H}$ do not exist. The space within these is "matter," without them "æther." All formulæ are to be deduced from the principle of least action. Following Larmor, suppose that the action, in so far as it involves $\mathbf{E}$ and $\mathbf{H}$, is identified with the expression

$$(8\pi)^{-1} \iiiint (H^2 - E^2)\, dv\, dt,$$

where dv is an element of volume, dt of time. It may be shown that the action is a minimum consistent with the conditions

$$\frac{\partial \mathbf{E}}{\partial t} = c\, \text{curl } \mathbf{H}, \quad \text{div } \mathbf{E} = 0, \quad \ldots\ldots\ldots\ldots(1)$$

if we have also

$$\frac{\partial \mathbf{H}}{\partial t} = -c\, \text{curl } \mathbf{E}, \quad \text{div } \mathbf{H} = 0; \quad \ldots\ldots\ldots(2)$$

(1) and (2) are the electromagnetic equations. And the

necessary conditions at any boundary of the electromagnetic field are just those which obtain at a perfect reflector.

Equations (1) and (2) possess integrals of different types. The quantity

$$\iint (\mathbf{E},\, d\mathbf{S})$$

is a constant, where $d\mathbf{S}$ is any element of area of a closed surface. Thus the electric induction over any closed surface is constant, and any portion of matter therefore carries a constant charge which is yet not an electric substance. Further, suppose the material surface is multiply-connected like an anchor ring. The magnetic induction across the aperture is constant, remembering that at the material surface the conditions are the same as those at a perfect reflector. Hence the anchor ring may behave like a permanent magnet of constant moment. The external field produces a tension of amount

$$(8\pi)^{-1}\,(E^2 - H^2)$$

per unit area of the surface of the matter. This tension accounts for all the observed magnetic and electric forces. Magnetic induction is now explicable, because this tension acting upon matter has replaced the force acting upon electric fluid. The molecular magnetic field is not to be explained as due to the circulation of electric current sheets. This is merely a mathematical device.

There remains the question what constraints must be applied to the material surfaces that they may be stable. That problem is not solved in any theory; it is no more or less soluble in the present theory than in any other.

THE MAGNETON AND PLANCK'S UNIVERSAL CONSTANT

[*From the Philosophical Magazine, Vol.* 26, 6th *Series, p.* 800.
October, 1913.]

To the Editors of the Philosophical Magazine.

GENTLEMEN,

In the discussion on the Theory of Radiation at the British Association Meeting just concluded, Dr Bohr's postulate of a natural unit of angular momentum was very prominent. The unit actually exists and is to be found in the magneton (see "A Theory of Gravity" in this number of your Magazine). A magneton of any cross-section or aperture has, I find, the angular momentum about its axis

$$(8\pi^2 V)^{-1} N_e N_\mu,$$

where V is the velocity of light, N_e is the number of tubes of electric induction terminating on the surface, N_μ the number of tubes of magnetic induction passing through the aperture. I hope with your permission to discuss in another number the meaning of this result for a theory of radiation and matter. By making Planck's h an angular momentum, Dr Bohr has introduced an idea of the first importance.

I am, Gentlemen,

Yours faithfully,

S. B. M^CLAREN.

UNIVERSITY COLLEGE, READING.

THE THEORY OF RADIATION

[*From Nature, Vol.* 92, *p.* 165. *Oct. 9th,* 1913.]

The natural unit of angular momentum postulated by Dr Niels Bohr, of Copenhagen, in his researches on the theory of spectral lines actually exists. It is the angular momentum of the magneton. Rejecting entirely the idea of magnetic or electric substance, the magneton may be regarded as an inner limiting surface of the æther, formed like an anchor ring. The tubes of electric induction which terminate on its surface give it an electric charge, the magnetic tubes linked through its aperture make it a permanent magnet.

I find that the angular momentum of any such system, whatever its shape or dimensions, about its axis of symmetry is

$$(8\pi^2 V)^{-1} \epsilon\mu,$$

where V is the velocity of light, ϵ the electric induction over the surface, and μ the magnetic induction over the aperture. I shall consider elsewhere the applications to the theory of complete radiation, spectral series, and the asymmetrical emission of electrons in ultra-violet light. Only this need be mentioned. If an electron (charge e) be thrown off from a magneton like a speck of dust from a flying wheel, then the angular momentum of the magneton changes by the amount $-e\mu (2\pi V)^{-1}$. This is therefore the angular momentum of the ejected electron about the axis of the magneton. Taking the velocity of ejection to be proportional to the angular velocity in the magneton, we have Ladenburg's result that the energy of the emitted rays varies as the frequency.

Dr Bohr, by first insisting on the fact that Planck's h is an angular momentum, has done something of the greatest importance, whatever the ultimate fate of his particular interpretation. Dr Nicholson has, I think, used the same idea.

S. B. MᶜLAREN.

UNIVERSITY COLLEGE, READING.
September 20.

A THEORY OF MAGNETS

[EDITORIAL NOTE: BY PROF. H. R. HASSÉ.

This paper is compiled from various manuscripts which were evidently intended to form two separate papers, one on "Magneton and Quantum" and the other on "A Theory of Magnets." The basis of the theory of both is to be found in the paper on "A Theory of Gravity" in the *Phil. Mag.*, Oct. 1913. As it is there bound up with a somewhat speculative theory of the gravitational field, McLaren intended to publish separately those portions which could be applied directly to a theory of magnetism, an abstract of which he gave to the meeting of the British Association in Sept. 1913. The results of immediate interest as regards this theory are given in § 2 of the paper below. Later on McLaren discovered that the magneton of his theory had the property that its angular momentum was constant, a result which appeared in the *Phil. Mag.* in October 1913. The proof of this, contained in the manuscript of the paper on "Magneton and Quantum," is given below in § 3. The next paragraph deals with the application of the Principle of Least Action to the case of a multiply connected surface like an anchor ring. As McLaren takes this principle as the basis of all theoretical investigations in the region of Mathematical Physics, it is essential to justify its use in this case. In the remaining paragraph the question of the dynamics of such a system for quasi-stationary motion is considered, and is treated on the lines given in McLaren's published papers.]

§ 1. *Introduction.*

It is claimed for Ampère that he was the first to give a real theory of Magnetism. This implies that Ampère resolved magnetic forces into something more fundamental when he showed that they were of electric origin. In terms of the modern electron theory of matter, that part of the force between two moving electric charges which depends on the velocities is identified with the magnetic force, and in this way Ampère is able to dispense with the old idea of polarised magnetic matter, used by Poisson. A permanent magnet is then accounted for by assuming that the electrons within the magnetised body describe closed orbits.

This theory, however, fails to account for the phenomenon of paramagnetism, as pointed out by W. Voigt*, nor does Langevin's theory in reality offer any explanation which goes deeper than Poisson's, where elementary magnets are already assumed. It is true that Langevin accounts for diamagnetism by resolving the elementary magnet into a swiftly circling electron, but when he comes to deal with paramagnetism he makes the fundamental assumption of all such theories. The electric current produces the same magnetic field as a certain magnetic doublet, therefore the reaction of an external field upon the current will produce the same effect as would be produced upon a permanent magnet.

This assumption cannot be justified, as is shown by the fact that the magnetic field acting upon a rotating electron can never do work, for the mechanical force is always directed at right angles to the motion of the electron. But if the electric current be replaced by a magnet, and the magnet is assumed to place itself parallel to the field, then certainly work is done. For a general argument leading to this result a previous paper of mine may be consulted†.

The purpose of this paper is to indicate the lines on which a theory can be built up which will not have the disadvantages pointed out above. Modern electron theory postulates the existence of a continuous substance "electricity" or "matter," and derives the electric and magnetic forces **E** and **H** from the motion of this matter, in accordance with the equations of the theory of electrons as given by Lorentz, viz.

$$\text{curl } \mathbf{H} = \frac{1}{c}\left(\frac{\partial \mathbf{E}}{\partial t} + 4\pi\rho\,\mathbf{u}\right), \quad \text{div } \mathbf{H} = 0$$

$$\text{curl } \mathbf{E} = -\frac{1}{c}\frac{\partial \mathbf{H}}{\partial t}, \qquad\qquad \text{div } \mathbf{E} = 4\pi\rho \right\} \quad \ldots\ldots(1)$$

---

* W. Voigt, *Ann. der Physik*, Vol. 9, p. 115, 1902.
† *Phil. Mag.*, Vol. 25, p. 55, Jan. 1913.

in which the density ρ and the velocity $\mathbf{u}$ of the electric charge are supposed known, and the electromagnetic field extends throughout all space, whether occupied by matter or not.

In the following pages, on the other hand, the fundamental equations are those of the æther alone, viz.

$$\left.\begin{array}{ll} \operatorname{curl}\mathbf{H} = \dfrac{1}{c}\dfrac{\partial\mathbf{E}}{\partial t}, & \operatorname{div}\mathbf{H} = 0 \\[2mm] \operatorname{curl}\mathbf{E} = -\dfrac{1}{c}\dfrac{\partial\mathbf{H}}{\partial t}, & \operatorname{div}\mathbf{E} = 0 \end{array}\right\}, \quad\ldots\ldots\ldots\ldots(2)$$

and matter is simply the boundary of the region to which these equations apply.

The electric charge is the integral of the normal component of $\mathbf{E}$ over any closed surface forming an internal boundary. Instead of a substance we have merely a constant of integration. When the boundary is multiply-connected—like an anchor ring —it will be shown that the flux of magnetic force over the aperture is also constant. Such a surface surrounded by a number of closed tubes of magnetic force is a permanent magnet —the magneton.

On this theory neither electrons nor magnetons appear as primary data, both are derived from the fundamental concept of the electromagnetic field. It is not stating the case unfairly if we say that whereas the electron theory involves both æther and matter, the theory of this paper employs only æther.

The regions occupied by matter are thus reduced to absolute vacua, all energy and momentum being electromagnetic and in the æther, not in matter. At the boundaries of matter and æther certain conditions have to be satisfied, which will be shown to be identical with those obtained by considering the boundary as a perfect reflector, and it follows then that the electromagnetic field produces a tension of definite amount per unit area on the surface of the matter.

This way of viewing the relations between matter and æther neglects however one most important consideration. The material surfaces must be supposed subject to constraints sufficient to secure their stability. This is the positive function of matter, and I do not myself believe that it can be regarded as merely mechanical. The constraints themselves cannot involve directly the possession of any energy, but by determining in part the forms of the material surfaces they determine also the distribution of electromagnetic energy.

Further it will be shown that the magneton possesses angular momentum of an amount

$$\frac{1}{8\pi c^2} N_e N_\mu,$$

where N_e is the number of tubes of electric force terminating on the surface, and N_μ the number of tubes of magnetic force passing through the aperture, a result which applies to any distribution of magnetons symmetrical about an axis and in a steady state. A comparison with the data supplied by Weiss' work on the magnetic moments of atoms shows that the magnetic moment per unit charge is of the same order as Planck's constant h. This supplies a basis in ordinary electromagnetic theory for the interpretation of h as an angular momentum, which we owe to Nicholson and Bohr, and angular momentum behaves as if atomic because charges are atomic. As a general principle for the deduction of the equations of the theory I shall assume the principle of least action, which I have used in previous papers*. As the work given in those papers is not sufficiently general to deal with the problem of multiply-connected surfaces, I shall give here the complete argument.

§ 2. Equations of the Electromagnetic Field.

The equations of the field are given in (2) above in their usual form, using Hertz-Heaviside units.

If $\dfrac{\partial'}{\partial t}$ denotes differentiation relative to axes moving with velocity u, so that

$$\frac{\partial' \mathbf{P}}{\partial t} = \frac{\partial \mathbf{P}}{\partial t} + u_x \frac{\partial \mathbf{P}}{\partial x} + u_y \frac{\partial \mathbf{P}}{\partial y} + u_z \frac{\partial \mathbf{P}}{\partial z},$$

---

* Phil. Mag., Oct. 1913, pp. 656–658. Proceedings of Mathematical Congress, Cambridge, 1912, Vol. II, p. 264.

where $\mathbf{P}$ is any vector, the equations of the field relative to the moving axes may be written

$$\operatorname{curl} \mathbf{H}' = \frac{1}{c}\frac{\partial' \mathbf{E}}{\partial t}, \qquad \operatorname{div} \mathbf{H} = 0 \left.\begin{array}{c}\\ \\ \\ \\ \end{array}\right\} , \dots\dots\dots\dots(3)$$
$$\operatorname{curl} \mathbf{E}' = -\frac{1}{c}\frac{\partial' \mathbf{H}}{\partial t}, \qquad \operatorname{div} \mathbf{E} = 0$$

where $\mathbf{E}'$ and $\mathbf{H}'$ are defined by

$$\mathbf{E}' = \mathbf{E} + \frac{1}{c}[\mathbf{u},\mathbf{H}], \quad \mathbf{H}' = \mathbf{H} - \frac{1}{c}[\mathbf{u},\mathbf{E}]. \quad \dots\dots(4)$$

The charge on a material surface is to be identified with the value of the quantity

$$(4\pi)^{-1}\int(\mathbf{E}, d\mathbf{S}),$$

where $d\mathbf{S}$ represents an element of the surface. That this is a constant follows from the first of equations (3), which may be written in the form

$$\frac{d}{dt}\left\{\int(\mathbf{E}, d\mathbf{S})\right\} = c\int(\operatorname{curl}\mathbf{H}', d\mathbf{S}) = c\int(\mathbf{H}', d\mathbf{s}), \dots(5)$$

where $\mathbf{s}$ is an element of arc of the curve bounding the surface S, and since the surface is in motion,

$$\frac{d}{dt}\left\{\int(\mathbf{E}, d\mathbf{S})\right\} = \int\left(\frac{\partial'\mathbf{E}}{\partial t}, d\mathbf{S}\right).$$

Since $\mathbf{E}$ and $\mathbf{H}$ are finite, single-valued, and continuous functions, it follows for a *closed* surface from (5) that

$$\frac{d}{dt}\left\{\int(\mathbf{E}, d\mathbf{S})\right\} = 0$$

or

$$\int(\mathbf{E}, d\mathbf{S})\dots\dots\dots\dots\dots\dots\dots(6)$$

is a constant, which we take as equal to $4\pi e$, where e is the electric charge on the surface.

As regards the conditions to be satisfied at a boundary between matter and æther, it is clear that, since $\mathbf{E}$ and $\mathbf{H}$ do not exist inside the matter, all the energy resides in the æther so that the boundary must act as a perfect reflector leading to the well-known conditions to be satisfied at a moving mirror, viz. that the normal component of **H** is zero, and the tangential component of **E'** is zero, conditions which may be deduced directly from the equations (3) in the usual manner*.

If now the surface is assumed to be doubly-connected like an anchor ring, it follows from the second of equations (3) that

$$\frac{d}{dt}\left\{\int(\mathbf{H}, d\mathbf{S})\right\} = -c\int(\mathbf{E'}, d\mathbf{s}), \quad \ldots\ldots\ldots\ldots(7)$$

where $d\mathbf{S}$ is now taken to be an element of area of a barrier drawn across the aperture of the ring, and $d\mathbf{s}$ an element of length of a closed curve lying in the material surface, and forming the boundary of the aperture. Since the tangential component of **E'** along the surface vanishes, it follows that

$$\int(\mathbf{H}, d\mathbf{S}) \quad \ldots\ldots\ldots\ldots\ldots\ldots(7\,a)$$

over any barrier across the aperture is constant, and equal to N_μ, the number of lines of magnetic force passing through the aperture.

It is well known that in the case of a fixed surface, the rate of change of electromagnetic momentum $\mathbf{M} = \dfrac{1}{4\pi c}\,[\mathbf{E}, \mathbf{H}]$ is equal to the total pressure on the surface due to the system of stresses as given by Maxwell, so that

$$\frac{\partial M_x}{\partial t} + \operatorname{div} P_x = 0, \quad \frac{\partial M_y}{\partial t} + \operatorname{div} P_y = 0, \quad \frac{\partial M_z}{\partial t} + \operatorname{div} P_z = 0,$$

where, using the notation of my paper "A Theory of Gravity," pp. 660–667,

$$P_x \equiv (P, U, T), \quad P_y \equiv (U, Q, S), \quad P_z \equiv (T, S, R).$$

In the case of a moving surface, these equations become

$$\frac{\partial' M_x}{\partial t} + \operatorname{div} P_x' = 0, \quad \frac{\partial' M_y}{\partial t} + \operatorname{div} P_y' = 0, \quad \frac{\partial' M_z}{\partial t} + \operatorname{div} P_z' = 0,$$

where P' represents the stress-system relative to the moving surface, if the principle of the conservation of electromagnetic momentum is to remain true.

---

* Cf. M. Abraham, *Theorie der Elektrizität*, Vol. ii, pp. 324, 325. 2nd Edition, 1905.

It will be found from the above equations that

$$P_x' = P_x - u_n M_x, \quad P_y' = P_y - u_n M_y, \quad P_z' = P_z - u_n M_z,$$

where u_n is the normal velocity of the moving surface, and in the case where the surface is a perfect reflector, the rate of change of momentum is normal to the surface and equal to

$$\frac{1}{8\pi}(H^2 - E^2)$$

per unit area*.

The same is true of the electromagnetic energy, so that the energy and the linear momentum increase exactly as if the material surfaces were acted on by a force of amount

$$\frac{1}{8\pi}(H^2 - E^2)$$

per unit area in the direction of the normal to the surface drawn into the electromagnetic field†.

For the sake of completeness I add a proof that the angular momentum also follows the same law. The angular momentum about the axis of z is, with the same notation,

$$\iiint (x M_y - y M_x)\, dv.$$

The rate of change of angular momentum of the field is therefore

$$\iiint \left(x\frac{\partial M_y}{\partial t} - y\frac{\partial M_x}{\partial t} \right) dv - \iint (x M_y - y M_x) u_n dS,$$

or $\qquad \iiint (y \operatorname{div} P_x - x \operatorname{div} P_y)\, dv - \iint (x M_y - y M_x) u_n dS,$

or $\quad \iiint \{\operatorname{div}(y P_x) - \operatorname{div}(x P_y)\}\, dv - \iint (x M_y - y M_x) u_n dS,$

or $\qquad \iint x (P_y - u_n M_y)\, dS - \iint y (P_x - u_n M_x)\, dS. \dots\text{(8)}$

Now $(P_x - u_n M_x)\, dS$ and $(P_y - u_n M_y)\, dS$ are the components of the pressure on the element dS of the moving surface resolved along the axes of x and y, and (8) is the moment about the axis

* For details of the calculation see p. 663 of paper quoted.
† For a direct calculation see Abraham, op. cit., pp. 329–333.

of z of the same pressure, so that the rate of change of angular momentum is equal to the moment of the pressures on the elements of the surface. [It will be seen that this paragraph gives the proof of the various statements in the abstract of the paper given to the British Association in 1913. Editor.]

§ 3. *The Angular Momentum of the Magneton.*

Consider any material system symmetrical about an axis. When the field is steady there exists an electrostatic potential ϕ, and a stream function ψ, the latter connected with the magnetic force. Take the axis of z along the axis of symmetry of the system, and suppose r is the distance of any point from this axis. Then the components of **E** are

$$-\frac{\partial \phi}{\partial z}, \quad -\frac{\partial \phi}{\partial r},$$

and those of **H** are

$$\frac{1}{r}\frac{\partial \psi}{\partial r}, \quad -\frac{1}{r}\frac{\partial \psi}{\partial z},$$

along and perpendicular to the axis.

The charge e on the surface is defined by

$$4\pi e = -\int \frac{\partial \phi}{\partial n} dS = -\int 2\pi \frac{\partial \phi}{\partial n} r\, ds,$$

where ds is an element of arc of the boundary on any section made by a plane through the axis of z.

The flux of magnetic force over any aperture of a multiply-connected surface is

$$\int (\mathbf{H}, d\mathbf{S}) = \int \frac{1}{r}\frac{\partial \psi}{\partial r} \cdot 2\pi r\, dr = 2\pi \psi_0,$$

where ψ_0 is the constant value of the stream function ψ on the surface itself.

The electromagnetic momentum, being $\dfrac{1}{4\pi c}[\mathbf{E}, \mathbf{H}]$ per unit volume, is in this case everywhere perpendicular to the axis of z, and is equal to

$$\frac{1}{4\pi c}\left(\frac{1}{r}\frac{\partial \psi}{\partial z}\frac{\partial \phi}{\partial z} + \frac{1}{r}\frac{\partial \psi}{\partial r}\frac{\partial \phi}{\partial r}\right) 2\pi r\, dr\, dz,$$

whence the moment of the momentum, or the angular momentum, is equal to

$$\frac{1}{2c} \int \left(r \frac{\partial \phi}{\partial z} \frac{\partial \psi}{\partial z} + r \frac{\partial \phi}{\partial r} \frac{\partial \psi}{\partial r} \right) dr\,dz,$$

which by partial integration with respect to z and r respectively becomes

$$-\frac{1}{2c} \int \psi r \frac{\partial \phi}{\partial n} . ds - \frac{1}{2c} \iint \psi \left\{ \frac{\partial}{\partial z} \left(r \frac{\partial \phi}{\partial z} \right) + \frac{\partial}{\partial r} \left(r \frac{\partial \phi}{\partial r} \right) \right\} dr\,dz.$$

The double integral vanishes since ϕ satisfies Laplace's equation

$$\frac{1}{r} \frac{\partial}{\partial r} \left(r \frac{\partial \phi}{\partial r} \right) + \frac{\partial^2 \phi}{\partial z^2} = 0,$$

and the angular momentum is therefore equal to

$$-\frac{1}{2c} \psi_0 \int r \frac{\partial \phi}{\partial n} \, ds$$

or to

$$-\frac{1}{2c} . \psi_0 (- 2e),$$

or, finally, to $\qquad \psi_0 e/c.$

Since the number N_μ of lines of magnetic force through the aperture is equal to $2\pi\psi_0$, and the number N_e of lines of electric force ending on the surface is $4\pi e$, the angular momentum may be written in the form

$$\frac{1}{8\pi^2 c} N_e N_\mu, \quad \dots\dots\dots\dots\dots\dots(9)$$

as given in my letter to *Phil. Mag.*, Vol. 26, p. 800. 1913. It will be noticed that the angular momentum is proportional to the charge e, so that if the angular momentum is atomic, it is due to the fact that the electric charge itself is atomic.

§ 4. *The Principle of Least Action for a Multiply-Connected Surface.*

The physical state of the æther requires for its definition four independent magnitudes. One of these is a scalar Ω, the others are the three components of a vector **A**. In terms of **A** and Ω the two vectors **E** and **H** are defined by the relations

$$\mathbf{H} = \frac{1}{c} \frac{\partial \mathbf{A}}{\partial t} - \nabla\Omega, \quad \mathbf{E} = \operatorname{curl} \mathbf{A}, \dots\dots\dots(10)$$

so that the two equations

$$\frac{\partial \mathbf{E}}{\partial t} = c \operatorname{curl} \mathbf{H}, \quad \operatorname{div} \mathbf{E} = 0$$

of the set of equations (2) are identically satisfied. It remains to deduce the other two equations of this set, the boundary conditions to be satisfied, and the equations of motion of the boundaries by varying the action $\mathscr{A}$ given by

$$\mathscr{A} = \frac{1}{8\pi} \iiiint (H^2 - E^2) \, dv \, dt. \quad \ldots\ldots\ldots\ldots(11)$$

From the result (6) of § 2, it follows that we must have

$$\iint (\operatorname{curl} \mathbf{A}, \, d\mathbf{S}) = 4\pi e,$$

which shows that the function $\mathbf{A}$ cannot be everywhere finite and single-valued. In the case of an anchor ring $\mathbf{A}$ is finite over the surface, but not single-valued. On completing a circuit round a circular section of the ring made by a plane through its axis the value of $\mathbf{A}$ does not return to its original value, behaving in this respect like the potential of a magnetic shell.

Suppose quite generally that the field external to matter is multiply-connected. The connectivity may be made simple by barriers placed across the apertures of any surfaces shaped like the anchor ring just taken as an example. The surface element dS is a vector element of area either of a barrier or of a material surface, but in the former case each element of area will occur twice in surface integrals. The vector element is such that the normal is drawn into the field from the surface. The barriers are within limits continuously deformable, and the differences in the final and initial values of $\mathbf{A}$ and Ω will be denoted by

$$(\mathbf{A})_0' \text{ and } (\Omega)_0'.$$

Since $\mathbf{E}$ and $\mathbf{H}$ are single-valued, it follows from (10) that

$$\operatorname{curl} (\mathbf{A})_0' = 0, \quad \frac{1}{c}\frac{d}{dt}(\mathbf{A})_0' - \nabla (\Omega)_0' = 0,$$

and therefore

$$(\mathbf{A})_0' = \nabla \chi, \quad (\Omega)_0' = \frac{1}{c}\frac{\partial \chi}{\partial t}. \quad \ldots\ldots\ldots\ldots(11\,a)$$

In calculating the variation of $\mathbf{A}$ in (11), I suppose not merely that the values of $\mathbf{E}$ and $\mathbf{H}$ at points within the field are varied, but also that the boundaries of the field, the material surfaces, are displaced. This is essential in order to obtain the equations of motion of these surfaces. We have then

$$\delta\mathscr{A}=(4\pi)^{-1}\iiiint \{(\mathbf{H},\,\delta\mathbf{H})-(\mathbf{E},\,\delta\mathbf{E})\}\,dv\,dt$$
$$-(8\pi)^{-1}\iiint (H^2-E^2)\,dn\,dS\,dt,\ldots(12)$$

where dn is the normal displacement of the surface element dS. This surface integral occurs obviously only at the material surfaces since $\mathbf{H}$ and $\mathbf{E}$ are single-valued, and at the barriers the terms on each side cancel.

If $\delta\mathbf{A}$ and $\delta\Omega$ are the variations in $\mathbf{A}$ and Ω, we have from (10)

$$\delta\mathbf{H}=\frac{1}{c}\frac{d}{dt}(\delta\mathbf{A})-\nabla(\delta\Omega),\quad \delta\mathbf{E}=\mathrm{curl}\,\delta\mathbf{A},$$

so that the first term $\delta\mathscr{A}_1$ of $\delta\mathscr{A}$ becomes

$$(4\pi)^{-1}\iiiint\left\{\left(\mathbf{H},\frac{1}{c}\frac{d}{dt}\delta\mathbf{A}\right)-(\mathbf{H},\nabla\delta\Omega)-(\mathbf{E},\,\mathrm{curl}\,\delta\mathbf{A})\right\}dv\,dt$$
$$\ldots\ldots\ldots(13)$$

or

$$\delta\mathscr{A}_1=(4\pi)^{-1}\iiiint\left\{\frac{1}{c}\frac{d}{dt}(\mathbf{H},\,\delta\mathbf{A})-\nabla(\mathbf{H}\delta\Omega-[\mathbf{E},\,\delta\mathbf{A}])\right.$$
$$\left.-\left(c^{-1}\frac{d}{dt}\mathbf{H}+\mathrm{curl}\,\mathbf{E},\,\delta\mathbf{A}\right)+\mathrm{div}\,\mathbf{H}\delta\Omega\right\}dv\,dt.\ \ldots(14)$$

The first two terms on the right-hand side of (14) are reducible to surface integrals, while the volume integrals vanish if

$$c^{-1}\frac{d}{dt}\mathbf{H}+\mathrm{curl}\,\mathbf{E}=0,\quad \mathrm{div}\,\mathbf{H}=0.$$

These are the remaining equations of the electromagnetic field, and (14) becomes

$$\delta\mathscr{A}_1=(4\pi)^{-1}\iiiint\left\{c^{-1}\frac{d}{dt}(\mathbf{H},\,\delta\mathbf{A})-\nabla(\mathbf{H}\delta\Omega-[\mathbf{E},\,\delta\mathbf{A}])\right\}dv\,dt.$$
$$\ldots\ldots\ldots(15)$$

Since

$$\iiiint c^{-1}\frac{d}{dt}(\mathbf{H},\,\delta\mathbf{A})\,dv\,dt=\int c^{-1}\frac{d}{dt}\left\{\iiint(\mathbf{H},\,\delta\mathbf{A})\,dv\right\}dt$$
$$-\iiint c^{-1}(\mathbf{H},\,\delta\mathbf{A})\,u_n\,dS\,dt,$$

where u_n is the velocity normal to itself of any element of the area of the boundary, (15) may be written

$$\delta\mathscr{A}_1 = (4\pi)^{-1} \int c^{-1} \frac{d}{dt} \left\{ \iiint (\mathbf{H}, \delta\mathbf{A})\, dv \right\} dt$$

$$- (4\pi)^{-1} \iiint c^{-1} (\mathbf{H}, \delta\mathbf{A})\, u_n dSdt$$

$$+ (4\pi)^{-1} \iiint (\mathbf{H}\delta\Omega - [\mathbf{E}, \delta\mathbf{A}], d\mathbf{S})\, dt. \ldots (16)$$

Where the integrals refer to the material surfaces, they may be made to vanish in (16) by assuming the normal component of $\mathbf{H}$ and the tangential component of $\mathbf{E}'$ to vanish. For since

$$\mathbf{E}' = \mathbf{E} + \frac{1}{c}\, [\mathbf{u}, \mathbf{H}], \quad \ldots\ldots\ldots\ldots\ldots (17)$$

$$\iint (\mathbf{H}\delta\Omega + [\mathbf{E}, \delta\mathbf{A}], d\mathbf{S}) = \iint \delta\Omega\, (\mathbf{H}, d\mathbf{S}) - \iint ([\mathbf{E}', \delta\mathbf{A}], d\mathbf{S})$$

$$+ c^{-1} \iint (d\mathbf{S}, [[\mathbf{u}, \mathbf{H}], \delta\mathbf{A}]),$$

and

$$- c^{-1} \iint (d\mathbf{S}, [[\mathbf{u}, \mathbf{H}], \delta\mathbf{A}])$$

$$= c^{-1} \iint \{ (d\mathbf{S}, \mathbf{u})\, (\mathbf{H}, \delta\mathbf{A}) - (d\mathbf{S}, \mathbf{H})\, (\mathbf{u}, \delta\mathbf{A}) \},$$

the second and third integrals in (16) may be written

$$\iiint \{ \delta\Omega + c^{-1} (\mathbf{u}, \delta\mathbf{A}) \}\, (\mathbf{H}, d\mathbf{S})\, dt + \iiint ([\mathbf{E}', \delta\mathbf{A}], d\mathbf{S})\, dt.$$
$$\ldots\ldots\ldots (18)$$

Where the integrals refer to the barriers, each element of area appears twice, and by (11 a) the expression (18) becomes

$$\iiint c^{-1} \frac{d'}{dt}\, (\delta\chi)\, (\mathbf{H}, d\mathbf{S})\, dt + \iiint ([\mathbf{E}', \nabla\delta\chi]\, d\mathbf{S})\, dt, \ldots (19)$$

where $\dfrac{d'}{dt}$ denotes the time rate of change at a point moving with the velocity of the barrier itself.

Consider now the integral

$$\iint \delta\chi\, (\mathbf{H}, d\mathbf{S})$$

extended over any barrier. Its time rate of change contains a term due to the fact that the velocity of the material surfaces is not the same as that of the barrier. Denote by ds an element of the curve of intersection of the barrier and the material surface. Let $\mathbf{v}$ be the velocity of the barrier relative to the material surface, then the area of the barrier decreases at the rate $[\mathbf{v}, \mathbf{ds}]$ owing to the motion of dS, and therefore

$$\frac{d}{dt}\left\{\iint \delta\chi\,(\mathbf{H}, d\mathbf{S})\right\} = \frac{d'}{dt}\left\{\iint \delta\chi\,(\mathbf{H}, d\mathbf{S})\right\} + \int \delta\chi\,([\mathbf{v}, \mathbf{ds}], \mathbf{H}),$$
$$\dots\dots(20)$$

where the symbol $\dfrac{d'}{dt}$ denotes as before the rate of change for points all of which move with the velocity of the barrier normal to itself.

From equation (7) it follows that

$$\frac{d'}{dt}(\mathbf{H}, d\mathbf{S}) = -\,c\,(\mathrm{curl}\,\mathbf{E}', d\mathbf{S}),$$

so that (20) becomes

$$\frac{d}{dt}\left\{\iint \delta\chi\,(\mathbf{H}, d\mathbf{S})\right\} = \iint \frac{d'}{dt}(\delta\chi)\,(\mathbf{H}, d\mathbf{S}) - c\iint \delta\chi\,(\mathrm{curl}\,\mathbf{E}', d\mathbf{S})$$
$$+ \int \delta\chi\,([\mathbf{v}, \mathbf{ds}], \mathbf{H}). \dots(21)$$

Now

$$\iint \delta\chi\,(\mathrm{curl}\,\mathbf{E}', d\mathbf{S}) = -\iint ([\mathbf{E}', \nabla\,\delta\chi]\,d\mathbf{S}) + \int \delta\chi\,(\mathbf{E}', d\mathbf{s}),$$
$$\dots\dots(22)$$

whence, substituting in (21) and rearranging,

$$\iint \frac{1}{c}\frac{d'}{dt}(\delta\chi)\,(\mathbf{H}, d\mathbf{S}) + \iint ([\mathbf{E}', \nabla\,\delta\chi], d\mathbf{S})$$
$$= \frac{1}{c}\frac{d}{dt}\left\{\iint \delta\chi\,(\mathbf{H}, d\mathbf{S})\right\} + \int \delta\chi\left\{(\mathbf{E}', d\mathbf{s}) - c^{-1}([\mathbf{v}, \mathbf{H}], d\mathbf{s})\right\}.$$
$$\dots\dots(23)$$

In (23) the value of $\mathbf{E}'$ as defined by (17) is calculated for a point moving with the velocity $(\mathbf{u}+\mathbf{v})$ of the barrier. If this value of $\mathbf{E}'$ is denoted by $\mathbf{E}''$, we have

$$(\mathbf{E}'', \mathbf{ds}) = (\mathbf{E}' + c^{-1}[\mathbf{v}, \mathbf{H}], d\mathbf{s}),$$

where $\mathbf{E}'$ refers to a point moving with the velocity of the material surface.

The former of these expressions is zero since the tangential component of $\mathbf{E}'$ is zero, whence (23) becomes

$$\frac{1}{c}\iint \frac{d'}{dt}(\delta\chi)\,(\mathbf{H}, d\mathbf{S}) + \iint([\mathbf{E}', \nabla\,\delta\chi]\,d\mathbf{S}) = c^{-1}\frac{d}{dt}\left\{\iint \delta\chi\,[\mathbf{H}, d\mathbf{S})\right\}.$$
$$\dots\dots\dots(24)$$

Thus the quantity in (19) is equivalent to the right-hand side of (24), so that, introducing it into (16) and (12), we find that

$$\delta\mathscr{A} = (4\pi c)^{-1}\int \frac{d}{dt}\left(\iiint(\mathbf{H}, \delta\mathbf{A})\,dv\right)dt$$
$$+ (4\pi c)^{-1}\int \frac{d}{dt}\left\{\iint \delta\chi\,(\mathbf{H}, d\mathbf{S})\right\}dt - (8\pi)^{-1}\iiint(H^2 - E^2)\,dn\,dS\,dt.$$
$$\dots\dots\dots(25)$$

The terms which are perfect differential coefficients with respect to the time being omitted, we have finally

$$\delta\mathscr{A} = -(8\pi)^{-1}\iiint(H^2 - E^2)\,dn\,dS\,dt,\dots\dots\dots(26)$$

so that $\mathscr{A}$ has a stationary value provided that for all displacements consistent with the material constraints

$$(8\pi)^{-1}\iint(H^2 - E^2)\,dn\,dS = 0. \dots\dots\dots\dots(27)$$

This equation includes implicitly the equations of motion of the matter, since, giving dn values consistent with the displacement and rotation of the material system as a single rigid body, it shows that the total force and total couple due to the pressure

$$(8\pi)^{-1}(H^2 - E^2)$$

vanish for all directions.

That this pressure leads to results consistent with the principles of conservation of energy, of linear and of angular momentum has been proved directly in § 2. There is one further point to be considered. The equation (22) assumes that $\delta\chi$ is a single-valued function over the surface of the barrier. This assumption is justified, its truth following from the fact that $\mathbf{E}$ is a single-valued function, and therefore that

$$\iint \text{curl} \,(\mathbf{A})_0{}' \, dS = 0, \quad \ldots\ldots\ldots\ldots(28)$$

where the integration is extended over any part of the area of a barrier. Hence if the variation of E is also single-valued,

$$\iint \text{curl} \,(\delta\mathbf{A})_0{}' \, dS = 0,$$

and therefore by (11 a),

$$\iint \text{curl} \nabla \,(\delta\chi) \, dS = 0 \ \text{ or } \ \int \nabla \,(\delta\chi) \, ds = 0,$$

where ds is an element of any closed circuit lying in the barrier, whence it follows that $\delta\chi$ is single-valued.

§ 5. *Quasi-stationary motion of the system.*

In order to determine the equations of motion of a material system it is necessary to make some assumption as to the degrees of freedom of each material surface. The simplest possible assumption, no longer legitimate for those who accept the theory of relativity, is that each piece of matter moves as a rigid body. For the case of an electron the formulæ derived on this basis are well known. If this procedure is followed in this case, it would be necessary to calculate the values of $\mathbf{E}$ and $\mathbf{H}$ in so far as they depend on the position, motion, and acceleration of the moving surfaces. This is a cumbrous process and involves a detailed knowledge of the way in which the elements of matter are arranged.

A more satisfactory method is to calculate the electromagnetic field on the assumption that it is at any instant sufficiently defined by the instantaneous positions and velocities of the material surfaces. It is true that this method can only be applied when the material velocities are small compared with the velocity of light, and when the radiation emitted is of wave-length large compared with the dimensions of the elementary molecular system. So far as we know this latter condition is always satisfied, and its truth involves in all probability the truth of the former statement as to the velocities of the material surface.

For the purpose of the calculation it is convenient to use Maxwell's vector potential $\mathbf{F}$ and scalar potential ϕ defined by

$$\mathbf{E} = -\frac{1}{c}\frac{\partial \mathbf{F}}{\partial t} - \nabla\phi, \quad \mathbf{H} = \operatorname{curl} \mathbf{F}.$$

If dr be any linear element lying in a material surface and moving in any way with it, we have

$$\frac{\partial'}{\partial t}(\mathbf{F}, d\mathbf{r}) = \left(\frac{\partial \mathbf{F}}{\partial t} + (\mathbf{u}, \nabla)\,\mathbf{F}, d\mathbf{r}\right) + (\mathbf{F}, d\mathbf{u}), \quad \dots(29)$$

where $\dfrac{\partial'}{\partial t}$ denotes as before differentiation with respect to the time at a point moving with velocity $\mathbf{u}$, and

$$(u, \nabla)\,\mathbf{F} = u_x\frac{\partial}{\partial x}\,\mathbf{F} + u_y\frac{\partial}{\partial y}\,\mathbf{F} + u_z\frac{\partial}{\partial z}\,\mathbf{F}$$

for any vector $\mathbf{F}$.

But

$$((\mathbf{u}, \nabla)\,\mathbf{F}, d\mathbf{r}) + (\mathbf{F}, d\mathbf{u}) = d\,\{(\mathbf{u}, \mathbf{F})\} - ([\mathbf{u}, \operatorname{curl} \mathbf{F}], d\mathbf{r}),$$

so that

$$\frac{\partial'}{\partial t}(\mathbf{F}, d\mathbf{r}) = \left(\frac{\partial \mathbf{F}}{\partial t} - [\mathbf{u}, \operatorname{curl} \mathbf{F}], d\mathbf{r}\right) + d\,\{(\mathbf{u}, \mathbf{F})\}. \dots(30)$$

Since, however,

$$\mathbf{E}' = -\frac{1}{c}\frac{\partial \mathbf{F}}{\partial t} - \nabla\phi + \frac{1}{c}[\mathbf{u}, \operatorname{curl} \mathbf{F}],$$

and since the tangential component of $\mathbf{E}'$ vanishes at the surface, (30) above becomes

$$\frac{\partial'}{\partial t}(\mathbf{F}, d\mathbf{r}) = d\,\{(\mathbf{u}, \mathbf{F}) - c\phi\}. \quad \dots\dots\dots\dots(31)$$

Further the normal component of $\mathbf{H}$ being zero at the surface, the tangential components of $\mathbf{F}$ must be of the form $\nabla\chi$, so that (31) may be written

$$d\left\{c\phi - (\mathbf{u}, \mathbf{F}) + \frac{1}{c}\frac{\partial'\chi}{\partial t}\right\} = 0,$$

whence
$$\phi - c^{-1}(\mathbf{u}, \mathbf{F}) + \frac{1}{c}\frac{\partial'\chi}{\partial t} = K, \dots\dots\dots\dots(32)$$

where K has the same value all over any one material surface.

The scalar and vector potentials ϕ and $\mathbf{F}$ are not uniquely defined in terms of $\mathbf{E}$ and $\mathbf{H}$, since we may add on to ϕ a term of the form $-\dfrac{1}{c}\dfrac{\partial \phi_0}{\partial t}$ if at the same time $\nabla\phi_0$ is added to $\mathbf{F}$. We shall suppose this function ϕ_0 so chosen that the normal component of $\mathbf{F}$ vanishes. If now the velocity u in (32) be always chosen to be the normal velocity of the material surface, this equation becomes

$$\phi = K - \frac{1}{c}\frac{\partial' \chi}{\partial t}. \quad \dots\dots\dots\dots\dots(33)$$

To determine the constant K we have the electrical charge on the surface

$$\iint \left(-\frac{1}{c}\frac{\partial \mathbf{F}}{\partial t} - \nabla\phi,\, d\mathbf{S}\right) = 4\pi e. \quad \dots\dots\dots\dots(34)$$

Since

$$\frac{\partial'}{\partial t}(\mathbf{F},\, d\mathbf{S}) = \left(\frac{\partial \mathbf{F}}{\partial t} + \mathbf{u}\operatorname{div}\mathbf{F},\, d\mathbf{S}\right) - (\operatorname{curl}[\mathbf{u},\mathbf{F}],\, d\mathbf{S}),$$

and since $(\mathbf{F},\, d\mathbf{S})$ is zero, it follows, by integrating over the whole surface, that

$$\iint \left(\frac{\partial \mathbf{F}}{\partial t} + \mathbf{u}\operatorname{div}\mathbf{F},\, d\mathbf{S}\right) = 0,$$

so that (34) becomes

$$-\iint (\nabla\phi - c^{-1}\mathbf{u}\operatorname{div}\mathbf{F},\, d\mathbf{S}) = 4\pi e. \quad \dots\dots\dots(35)$$

Again, it is always possible to choose ψ_0 so that $\operatorname{div}\mathbf{F} = 0$, and this can be done without changing the normal component of $\mathbf{F}$, which remains zero. For if $\nabla\psi_0$ be added to $\mathbf{F}$, giving $\mathbf{F}'$, we can make $\operatorname{div}\mathbf{F}'$ zero provided

$$\nabla^2\psi_0 + \operatorname{div}\mathbf{F} = 0,$$

and ψ_0 may be chosen to satisfy this equation and at the same time the normal component of $\nabla\psi_0$ may have any assigned value, in this particular case given by making $\dfrac{\partial\psi_0}{\partial n} + \mathbf{F}_n$ zero.

With this choice of ψ_0, (35) becomes

$$-\iint (\nabla\phi,\, d\mathbf{S}) = 4\pi e. \quad\dots\dots\dots\dots\dots(36)$$

Finally, the differential equations satisfied by ϕ and $\mathbf{F}$ are

$$\left.\begin{aligned}\left(\nabla^2-\frac{1}{c^2}\frac{\partial^2}{\partial t^2}\right)\mathbf{F}-\nabla\left(\operatorname{div}\mathbf{F}+\frac{1}{c}\frac{\partial\phi}{\partial t}\right)=0\\\left(\nabla^2-\frac{1}{c^2}\frac{\partial^2}{\partial t^2}\right)\phi+\frac{1}{c}\frac{\partial}{\partial t}\left(\operatorname{div}\mathbf{F}+\frac{1}{c}\frac{\partial\phi}{\partial t}\right)=0\end{aligned}\right\}\dots\dots(37)$$

When div $\mathbf{F}$ is zero, these become

$$\left.\begin{aligned}\left(\nabla^2-\frac{1}{c^2}\frac{\partial^2}{\partial t^2}\right)\mathbf{F}-\nabla\left(\frac{1}{c}\frac{\partial\phi}{\partial t}\right)=0\\\nabla^2\phi=0\end{aligned}\right\}\dots\dots\dots(38)$$

The surface conditions to be satisfied are that the tangential component of $\mathbf{F}$ is $\nabla\chi$ and the normal component is zero. In addition there is the condition given by (33), where the constant K is to be determined from (36).

These formulæ are sufficient to determine $\mathbf{F}$ and ϕ as functions of the material state.

MATHEMATICAL THEORY OF THE MAGNETON

[EDITORIAL NOTE: BY PROF. H. R. HASSÉ.

The investigations given in the following pages are compiled from about eighty pages of manuscript, almost entirely mathematical. These deal largely with the electromagnetic field due to an electrically charged anchor ring (magneton) of small circular section together with the electric field due to a point charge (electron) at its centre. It has not been possible to do more than check the analytical results, and to reproduce them in the hope that they may be found to have a definite physical application.

§§ 1–5 of the following paper deal with the mathematical analysis of the field, which is a necessary preliminary to any further work, and follow closely the somewhat analogous analysis of the hydrodynamical vortex theory.

§ 6 deals with the question of the oscillations of the central electron under the influence of the field of the magneton, and is probably introductory to more general problems of a similar nature.

§ 7 deals with the difficult question of the numerical magnitudes of the various quantities in physical applications, and here it is impossible to do more than suggest the lines on which McLaren may have intended to proceed. As the manuscript gives no indication of any final result, this paragraph is entirely due to the editor of this section, who must be held responsible for the statements contained in it.]

§ 1. *Preliminary mathematical analysis.*

The simplest case of a multiply-connected surface is given by an anchor ring, so that the magneton will be represented by such a ring of (small) circular section lying in the plane through the origin of coordinates perpendicular to the axis of z, the centre of the ring being at the origin. The field due to such a ring is symmetrical about its axis, and therefore expressible in terms of coordinates z and ρ, the latter being the distance of any point from the axis of z.

It is necessary to express the differential equations of the theory in terms of the curvilinear coordinates ξ, η defined by the equation

$$z + i\rho = a \tan \tfrac{1}{2} (\xi + i\eta), \quad \dots\dots\dots\dots(1)$$

from which $\quad z = \dfrac{a \sin \xi}{\cosh \eta + \cos \xi}, \quad \rho = \dfrac{a \sinh \eta}{\cosh \eta + \cos \xi}, \dots\dots\dots(2)$

and the distance ds between neighbouring points is given by

$$(ds)^2 = J^2 \{(d\xi)^2 + (d\eta)^2\},$$

where $\quad J^2 = \left(\dfrac{\partial z}{\partial \xi}\right)^2 + \left(\dfrac{\partial z}{\partial \eta}\right)^2 = \left(\dfrac{\partial \rho}{\partial \xi}\right)^2 + \left(\dfrac{\partial \rho}{\partial \eta}\right)^2,$

so that $\quad J = a/(\cosh \eta + \cos \xi). \quad \dots\dots\dots\dots(3)$

In three dimensions the surfaces $\eta = $ constant given by the equation

$$z^2 + \rho^2 - 2a\rho \coth \eta + a^2 = 0$$

represent the anchor ring whose section is a circle of radius $a \operatorname{cosec} \eta$, the centre of which is at a distance $a \coth \eta$ from the axis of z. As $\eta \to \infty$ the anchor ring tends to the circle $\rho = a$, $z = 0$.

The conjugate surfaces $\xi = $ constant are portions of spheres having this latter circle as a plane section, those portions for which ξ lies between 0 and π being on the positive side of the plane $z = 0$, and those for which ξ lies between $-\pi$ and 0 being on the negative side. The surface for which $\xi = 0$ coincides with the disc $\rho = a$, $z = 0$.

Suppose now that X and Y represent the electric (or magnetic) force at any point P perpendicular respectively to the surfaces $\xi = $ const. and $\eta = $ const., which pass through P, their directions being those in which ξ and η increase. In the case of symmetry about the axis of z, the equations curl $\mathbf{E} = 0$, div $\mathbf{E} = 0$ reduce to

$$\left.\begin{aligned} \frac{\partial}{\partial \xi}(YJ) - \frac{\partial}{\partial \eta}(XJ) &= 0 \\[2mm] \frac{\partial}{\partial \xi}(XJ\rho) + \frac{\partial}{\partial \eta}(YJ\rho) &= 0 \end{aligned}\right\} \quad \dots\dots\dots\dots(4)$$

It therefore follows that we may take

$$YJ = \frac{\partial \phi}{\partial \eta}, \quad XJ = \frac{\partial \phi}{\partial \xi}, \quad \dots\dots\dots\dots(5)$$

and that ϕ satisfies the equation

$$\frac{\partial}{\partial \xi}\left(\rho \frac{\partial \phi}{\partial \xi}\right) + \frac{\partial}{\partial \eta}\left(\rho \frac{\partial \phi}{\partial \eta}\right) = 0, \quad \dots\dots\dots\dots(6)$$

which is of course the equation $\nabla^2 \phi = 0$ transformed into the curvilinear coordinates ξ and η.

Equations (4) may also be solved by taking

$$XJ\rho = -\frac{\partial \psi}{\partial \eta}, \quad YJ\rho = \frac{\partial \psi}{\partial \xi}, \quad \dots\dots\dots\dots(7)$$

so that ψ satisfies the equation

$$\frac{\partial}{\partial \xi}\left(\frac{1}{\rho}\frac{\partial \psi}{\partial \xi}\right) + \frac{\partial}{\partial \eta}\left(\frac{1}{\rho}\frac{\partial \psi}{\partial \eta}\right) = 0, \quad \dots\dots\dots\dots(8)$$

in which case ψ is Stokes' stream function, and (8) is equivalent to the transformation of the equation

$$\frac{\partial^2 \psi}{\partial z^2} + \frac{\partial^2 \psi}{\partial \rho^2} - \frac{1}{\rho}\frac{\partial \psi}{\partial \rho} = 0,$$

satisfied by ψ, into the coordinates ξ and η.

The solution of equations (6) and (8) involves the use of the toroidal functions introduced by Hicks, and studied in connection with analogous hydrodynamical investigations. The notation used in these pages is that of Basset, *Hydrodynamics*, Vol. II, particularly pp. 21–31, to which frequent reference will be made (in the form B, p. ...), as the properties of these functions needed for the present purpose are either already known or are easy deductions from known results.

The functions $P_n(z)$, $Q_n(z)$ required are the two independent solutions of the equation

$$\frac{d}{dz}\left\{(1 - z^2)\frac{d\chi_n}{dz}\right\} + (n^2 - \tfrac{1}{4})\chi_n = 0, \dots\dots\dots\dots(9)$$

where z is now used as an abbreviation for $\cosh \eta$.

$P_n(z)$ is that solution of (9) which is finite when $z = 1$ or $\eta = 0$, and infinite when $z = \infty$ or $\eta = \infty$, while $Q_n(z)$ is infinite when $z = 1$ and zero when $z = \infty$.

Equation (6) is a special case of Neumann's transformation (B, p. 8), and the solution (B, p. 22) can be expressed in the form

$$(\cosh \eta + \cos \xi)^{\frac{1}{2}} \, \Sigma \, \chi_n \cos (n\xi + a_n), \quad \ldots\ldots\ldots(10)$$

where the summation applies to positive integral values of n, including zero.

Equation (8) can be expressed in terms of an equation of the same type (B, p. 28), and the solution can be written

$$(\cosh \eta + \cos \xi)^{-\frac{1}{2}} \, \Sigma \, \sinh \eta \, \frac{d\chi_n}{d\eta} \cos (n\xi + a_n). \quad \ldots(11)$$

§ 2. *Electric field due to a stationary magneton, plus an electron at its centre.*

The condition that the tangential component of the electric force **E** is zero over the surface $\eta = \eta_0$ shows from (5) that $X = 0$, or ϕ is independent of ξ over the surface, so that the problem is that of the determination of a function ϕ satisfying equation (6) and due to a distribution of electricity on the surface of the anchor ring together with a point-charge at the origin, and such that ϕ has a constant value ϕ_0 over the surface $\eta = \eta_0$.

The value of ϕ due to the electrification of the anchor ring itself is given by

$$(\cosh \eta + \cos \xi)^{\frac{1}{2}} \, \overset{\infty}{\underset{0}{\Sigma}} \, a_n \, P_n (z) \cos n\xi,$$

and that due to a point-charge of amount e at the origin of coordinates is (B, p. 27)

$$\frac{e \sqrt{2}}{a\pi} (\cosh \eta + \cos \xi)^{\frac{1}{2}} \left(Q_0 + 2 \overset{\infty}{\underset{1}{\Sigma}} Q_n \cos n\xi \right).$$

The total potential ϕ is therefore equal to

$$(\cosh \eta + \cos \xi)^{\frac{1}{2}} \left\{ \overset{\infty}{\underset{0}{\Sigma}} (a_n P_n \cos n\xi) + \frac{e \sqrt{2}}{a\pi} \left(Q_0 + 2 \overset{\infty}{\underset{1}{\Sigma}} Q_n \cos n\xi \right) \right\}.$$
$$\ldots\ldots\ldots(12)$$

If this is constant over the surface $\eta = \eta_0$ of the anchor ring, we must have

$$\phi_0 (\cosh \eta + \cos \xi)^{-\frac{1}{2}} = \left(a_0 P_0 + \frac{e \sqrt{2}}{a\pi} Q_0 \right)$$
$$+ \overset{\infty}{\underset{1}{\Sigma}} \left(a_n P_n + \frac{2e \sqrt{2}}{a\pi} Q_n \right) \cos n\xi \ldots(13)$$

for $\eta = \eta_0$, and for all values of ξ.

Now (B, p. 26)

$$(\cosh \eta - \cos \xi)^{-\frac{1}{2}} = \frac{\sqrt{2}}{\pi}\left(Q_0 + 2\sum_1^\infty Q_n \cos n\xi\right),$$

whence, changing ξ into $\pi - \xi$,

$$(\cosh \eta + \cos \xi)^{-\frac{1}{2}} = \frac{\sqrt{2}}{\pi}\left(Q_0 + 2\sum_1^\infty (-)^n Q_n \cos n\xi\right),\dots(14)$$

so that, substituting in (13),

$$\phi_0 \frac{\sqrt{2}}{\pi}\left(Q_0 + 2\sum_1^\infty (-)^n Q_n \cos n\xi\right) = \left(a_0 P_0 + \frac{e\sqrt{2}}{a\pi}Q_0\right)$$
$$+ \sum_1^\infty\left(a_n P_n + \frac{2e\sqrt{2}}{a\pi}Q_n\right)\cos n\xi.$$

Equating coefficients of $\cos n\xi$ on both sides of this equation, the coefficients a_0 and a_n are determined by

$$\left.\begin{aligned} a_0 &= \frac{\sqrt{2}}{\pi}\frac{Q_0}{P_0}\left(\phi_0 - \frac{e}{a}\right) \\ a_n &= \frac{2\sqrt{2}}{\pi}\frac{Q_n}{P_n}\left((-)^n \phi_0 - \frac{e}{a}\right) \end{aligned}\right\},\dots\dots\dots(15)$$

where in the functions P_n and Q_n we must put $\eta = \eta_0$, or $z = z_0$. The total charge e_m on the magneton has now to be calculated, and is given by the equation

$$4\pi e_m = \iint Y dS = 2\pi \int \frac{1}{J}\frac{\partial\phi}{\partial\eta}\rho ds,$$

where s is the arc of the circular section.

Since $ds = J d\xi$, the equation becomes

$$2e_m = \int_{-\pi}^{+\pi}\rho\frac{\partial\phi}{\partial\eta}d\xi = \int_{-\pi}^{\pi}\rho\sinh\eta\frac{\partial\phi}{\partial z}d\xi = 2a\int_0^\pi\frac{\sinh^2\eta}{z+\cos\xi}\frac{\partial\phi}{\partial z}d\xi,$$

where $z = \cosh\eta$, and η is to be put equal to η_0 after differentiation.

From (12) we have

$$\frac{\partial\phi}{\partial z} = \frac{1}{2}(z+\cos\xi)^{-\frac{1}{2}}\left\{\left(a_0 P_0 + \frac{e\sqrt{2}}{a\pi}Q_0\right) + \sum_1^\infty\left(a_n P_n + \frac{2e\sqrt{2}}{a\pi}Q_n\right)\cos n\xi\right\}$$
$$+ (z+\cos\xi)^{\frac{1}{2}}\left\{\left(a_0\frac{dP_0}{dz} + \frac{e\sqrt{2}}{a\pi}\frac{dQ_0}{dz}\right)\right.$$
$$\left. + \sum_1^\infty\left(a_n\frac{dP_n}{dz} + \frac{2e\sqrt{2}}{a\pi}\frac{dQ_n}{dz}\right)\cos n\xi\right\}.$$

The first line of this last equation is from (12) equal to $\frac{1}{2}\phi_0/(z+\cos\xi)$, and the coefficient of $\cos n\xi$ in the second line is, using (15), equal to

$$a_n\frac{dP_n}{dz}+\frac{2e\sqrt{2}}{a\pi}\frac{dQ_n}{dz}=\frac{2\sqrt{2}}{\pi}\frac{Q_n}{P_n}\left[(-)^n\,\phi_0-\frac{e}{a}\right]\frac{dP_n}{dz}+\frac{2\sqrt{2}}{\pi}\frac{e}{a}\frac{dQ_n}{dz}$$

$$=\frac{2\sqrt{2}\,(-)^n\,\phi_0}{\pi}\frac{Q_n}{P_n}\frac{dP_n}{dz}+\frac{2\sqrt{2}}{\pi}\frac{e}{a}\cdot\frac{1}{P_n}\left\{P_n\frac{dQ_n}{dz}-Q_n\frac{dP_n}{dz}\right\}$$

$$=\frac{2\sqrt{2}\,(-)^n\,\phi_0}{\pi}\left(\frac{P_n\dfrac{dQ_n}{dz}+\dfrac{\pi}{S^2}}{P_n}\right)+\frac{2\sqrt{2}}{\pi P_n}\frac{e}{a}\left(-\frac{\pi}{S^2}\right),$$

[where $S\equiv\sinh\eta$, and $Q_n\dfrac{dP_n}{dz}-P_n\dfrac{dQ_n}{dz}=\dfrac{\pi}{S^2}$, (B, p. 27)]

$$=\frac{2\sqrt{2}\,(-)^n\,\phi_0}{\pi}\frac{dQ_n}{dz}+\frac{2\sqrt{2}}{S^2 P_n}[-e/a+(-)^n\,\phi_0].\ \ \ldots\ldots(16)$$

Similarly the term $a_0\dfrac{dP_0}{dz}+\dfrac{e\sqrt{2}}{a\pi}\dfrac{dQ_0}{dz}$ is equal to

$$\frac{\sqrt{2}}{\pi}\phi_0\frac{dQ_0}{dz}+\frac{\sqrt{2}}{P_0 S^2}(\phi_0-e/a),\ldots\ldots\ldots\ldots\ldots(17)$$

so that we have finally

$$\frac{\partial\phi}{\partial z}=\frac{1}{2}\phi_0/(z+\cos\xi)+(z+\cos\xi)^{\frac{1}{2}}\frac{\sqrt{2}\phi_0}{\pi}\left\{\frac{dQ_0}{dz}+2\Sigma\,(-)^n\frac{dQ_n}{dz}\cos n\xi\right\}$$

$$+(z+\cos\xi)^{\frac{1}{2}}\frac{\sqrt{2}}{S^2}\left\{\frac{\phi_0-e/a}{P_0}+2\Sigma\cos n\xi\frac{(-e/a+(-)^n\,\phi_0)}{P_n}\right\}.$$
$$\ldots\ldots(18)$$

By differentiating (14) with respect to z, it follows directly that

$$\frac{dQ_0}{dz}+2\overset{n}{\underset{1}{\Sigma}}(-)^n\frac{dQ_n}{dz}\cos n\xi=-\frac{1}{2}\frac{\pi}{\sqrt{2}}(z+\cos\xi)^{-\frac{3}{2}},\ldots(19)$$

so that the first two terms in (18) cancel, and we are left with

$$\frac{\partial\phi}{\partial z}=(z+\cos\xi)^{\frac{1}{2}}\frac{\sqrt{2}}{S^2}\left\{\frac{\phi_0-e/a}{P_0}+2\Sigma\cos n\xi\frac{(-e/a+(-)^n\,\phi_0)}{P_n}\right\}.$$
$$\ldots\ldots(20)$$

Finally, therefore, the charge e_m is given by

$$e_m=a\int_0^\pi\sqrt{2}\,(z+\cos\xi)^{-\frac{1}{2}}\left\{\frac{\phi_0-e/a}{P_0}+2\Sigma\cos n\xi\frac{(-e/a+(-)^n\,\phi_0)}{P_n}\right\}d\xi\Big\},$$

or, substituting for $(z + \cos \xi)^{-\frac{1}{2}}$ from (14),

$$e_m = \frac{2a}{\pi} \int_0^\pi [Q_0 + 2\Sigma (-)^n Q_n \cos n\xi] \left[\frac{\phi_0 - e/a}{P_0} \right.$$

$$\left. + 2\Sigma \cos n\xi \frac{(-e/a + (-)^n \phi_0)}{P_n} \right] d\xi$$

$$= 2a \left\{ \frac{Q_0}{P_0} (\phi_0 - e/a) + 2 \sum_1^\infty \frac{Q_n}{P_n} (\phi_0 - (-)^n e/a) \right\} \ldots\ldots\ldots(21)$$

This last result is exact, but, as will appear later, it is only possible to use the first term in the series, which is really a series in powers of k, $= e^{-\eta_0}$, so that k is small if the radius of the section of the ring is small. In this case $Q_0/P_0 = \pi/2L$ (B, p. 31), where $L = \log(4/k) = \log(8a/r)$, r being the radius of the section of the ring.

We then have $$e_m = \frac{a\pi}{L} (\phi_0 - e/a) \ldots\ldots\ldots\ldots\ldots(22)$$

as the important result of this paragraph.

§ 3. *Magnetic field due to a stationary magneton.*

The magnetic field has now to be determined from the equations

$$\text{curl } \mathbf{H} = 0, \qquad \text{div } \mathbf{H} = 0,$$

which lead to the same differential equations as in § 1, where X and Y are now the components of the magnetic force. In order then to obtain a solution of a different type to that in § 2, we must express X and Y in terms of the function ψ as in (7), leading to the expression for ψ given in (11). The condition that the normal component of $\mathbf{H}$ is to be zero makes Y vanish over the surface of the ring, or ψ is to be constant $(= \psi_0)$ for $\eta = \eta_0$.

The value of ψ appropriate to the problem for the space outside the ring is then given from (11) by

$$(z + \cos \xi)^{-\frac{1}{2}} (z^2 - 1) \Sigma b_n \frac{dP_n}{dz} \cos n\xi, \ldots\ldots\ldots(23)$$

where $z = \cosh \eta$, and we have replaced $\dfrac{dP_n}{d\eta}$ by $\sinh \eta \dfrac{dP_n}{dz}$.

The surface condition is then satisfied by determining the values of b_n, so that the equation

$$\psi_0 (z + \cos \xi)^{\frac{1}{2}} = (z^2 - 1) \overset{\infty}{\underset{0}{\Sigma}} \left(b_n \frac{dP_n}{dz} \cos n\xi \right) \quad \ldots (23\,a)$$

is true for all values of ξ, to do which we have to expand $(z + \cos \xi)^{\frac{1}{2}}$ in a Fourier's series in cosines, viz.

$$(z + \cos \xi)^{\frac{1}{2}} = \overset{\infty}{\underset{0}{\Sigma}} a_n \cos n\xi. \quad \ldots\ldots\ldots\ldots(24)$$

The coefficient a_n is therefore found from the equation

$$\frac{\pi}{2} a_n = \int_0^\pi (z + \cos \xi)^{\frac{1}{2}} \cos n\xi \, d\xi = (-)^n \int_0^\pi (z - \cos \xi)^{\frac{1}{2}} \cos n\xi \, d\xi$$

$$= (-)^n \frac{2\sqrt{2}}{4n^2 - 1} S \frac{dQ_n}{d\eta} = \frac{2\sqrt{2}}{4n^2 - 1} (-)^n S^2 \frac{dQ_n}{dz} \quad \ldots\ldots(24\,a)$$

[see B, p. 29], and a_0 is similarly given by

$$\pi a_0 = S^2 \frac{dQ_0}{dz} \cdot \frac{2\sqrt{2}}{0^2 - 1}. \quad \ldots\ldots\ldots\ldots(24\,b)$$

Hence (23) becomes

$$\psi_0 \frac{2\sqrt{2}}{\pi} S^2 \left\{ \frac{1}{0^2 - 1} \frac{dQ_0}{dz} + \overset{\infty}{\underset{1}{\Sigma}} \frac{2(-)^n}{4n^2 - 1} \frac{dQ_n}{dz} \cos n\xi \right\}$$

$$= (z^2 - 1) \left\{ b_0 \frac{dP_0}{dz} + \overset{\infty}{\underset{1}{\Sigma}} b_n \frac{dP_n}{dz} \cos n\xi \right\},$$

so that

$$\left. \begin{aligned} b_0 \frac{dP_0}{dz} &= \frac{\psi_0}{\pi \sqrt{2}} \frac{dQ_0}{dz} \Big/ (0^2 - \tfrac{1}{4}), \\ b_n \frac{dP_n}{dz} &= \frac{2\psi_0}{\pi \sqrt{2}} (-)^n \frac{dQ_n}{dz} \Big/ (n^2 - \tfrac{1}{4}) \end{aligned} \right\} \ldots\ldots(25)$$

in which z has to be replaced by z_0.

It may easily be verified that, in accordance with the general result given in the "Theory of Magnets" paper, the flux of magnetic force through the aperture of the ring is constant, being equal to

$$\int X 2\pi\rho \, ds = \int 2\pi\rho \left(\frac{1}{\rho J} \frac{\partial \psi}{\partial \eta} \right) J d\eta = 2\pi\psi_0, \ldots\ldots(26)$$

where ds is in fact the arc of the generating circle of any of the conjugate surfaces $\xi = $ constant, and therefore equal to $J d\eta$, while X is the magnetic force perpendicular to this surface.

There is a further constant of importance associated with the magnetic field of the magneton. It corresponds to the cyclic constant in the analogous hydrodynamical problem, which is the circulation round any closed curve threading the aperture once, and is calculated in Basset, p. 30.

In this case we have to calculate the integral of the magnetic force along the closed curve, which we will take to be a circle $\eta = $ constant, so that the integral is equal to

$$\int_{-\pi}^{+\pi} X ds = \int_{-\pi}^{+\pi} \left(\frac{1}{\rho J} \frac{\partial \psi}{\partial \eta} \right) J d\xi = \int_{-\pi}^{+\pi} \frac{1}{\rho} \frac{\partial \psi}{\partial \eta} d\xi$$

$$= \frac{1}{a} \int_{-\pi}^{+\pi} S^{-1} (\cosh \eta + \cos \xi) \frac{\partial \psi}{\partial \eta} d\xi = \frac{1}{a} \int_{-\pi}^{+\pi} (\cosh \eta + \cos \xi) \frac{\partial \psi}{\partial z} d\xi.$$

From (23) $\dfrac{\partial \psi}{\partial z}$ is equal to

$$-\frac{1}{2} (z + \cos \xi)^{-\frac{3}{2}} (z^2 - 1) \Sigma b_n \frac{dP_n}{dz} \cos n\xi$$

$$+ (z + \cos \xi)^{-\frac{1}{2}} \Sigma (n^2 - \tfrac{1}{4}) b_n P_n \cos n\xi, \dots (26\,a)$$

since $\qquad \dfrac{d}{dz} \left\{ (z^2 - 1) \dfrac{dP_n}{dz} \right\} = (n^2 - \tfrac{1}{4}) P_n,$

so that the circulation is equal to

$$\frac{2}{a} \int_0^\pi \left\{ -\frac{1}{2} (z^2 - 1)(z + \cos \xi)^{-\frac{1}{2}} \Sigma \left(b_n \frac{dP_n}{dz} \cos n\xi \right) \right.$$

$$\left. + (z + \cos \xi)^{\frac{1}{2}} \Sigma (n^2 - \tfrac{1}{4}) b_n P_n \cos n\xi \right\} d\xi. \dots (27)$$

Substituting for $(z + \cos \xi)^{-\frac{1}{2}}$ from (14) and for $(z + \cos \xi)^{\frac{1}{2}}$ from (24), and picking out the coefficient of b_n in (27), this term contributes to the circulation the amount

$$\frac{2}{a} b_n \cdot \frac{\pi}{2} \left\{ -\frac{1}{2} (z^2 - 1) \frac{dP_n}{dz} \frac{2\sqrt{2}}{\pi} (-)^n Q_n + (n^2 - \tfrac{1}{4}) P_n \cdot a_n \right\},$$

where a_n is given by (24 a).

The contribution of this term is therefore equal to

$$\frac{b_n}{a} \pi \left\{ -(z^2 - 1) \frac{dP_n}{dz} Q_n (-)^n \frac{\sqrt{2}}{\pi} + \frac{\sqrt{2}}{\pi} (-)^n (z^2 - 1) P_n \frac{dQ_n}{dz} \right\}$$

$$= -\frac{b_n \sqrt{2}}{a} (-)^n \left\{ (z^2 - 1) \left(\frac{dP_n}{dz} Q_n - P_n \frac{dQ_n}{dz} \right) \right\}$$

$$= -\frac{b_n \sqrt{2}}{a} (-)^n \pi. \quad \text{[Basset, p. 27, Equation (63).]}$$

The term independent of ξ contributes in the same way the amount $-b_0 \dfrac{\sqrt{2}}{a}\pi$ to the circulation, so that the total circulation is

$$-\frac{\pi\sqrt{2}}{a}\Sigma(-)^n b_n,\dots\dots\dots\dots\dots(28)$$

which to a first approximation is equal to

$$-\frac{\pi\sqrt{2}}{a}b_0 = -\frac{\pi\sqrt{2}}{a}\cdot\frac{\psi_0}{\pi\sqrt{2}}\left\{-4\frac{dQ_0}{dz}\Big/\left(\frac{dP_0}{dz}\right)\right\}$$

$$=\frac{4\psi_0}{a}\cdot\frac{\pi}{2(L-2)}=2\pi\psi_0/a\,(L-2),\,\dots(29)$$

using the series in k given by Basset, p. 31.

To this order of approximation the complete circulation through the ring is obtained by multiplying the above by the area πa^2 of the aperture, giving

$$2\pi^2 a\psi_0/(L-2)$$

as the value of the complete circulation.

Since the circulation is analogous to the work done in moving a magnetic pole once round a closed curve threading a current circuit, and therefore equal to 4π times the current, the magnetic moment of the magneton is to the same order of approximation equal to

$$\pi a^2\,(\text{current}) = 4\pi\,\frac{2\pi^2 a\psi_0}{(L-2)}$$

$$=\pi a\psi_0/2\,(L-2).\,\dots\dots\dots(30)$$

This result can be checked by working out the magnetic force due to the magneton at a great distance, which gives an expression for the magnetic moment.

§ 4. *The electromagnetic pressure on the magneton.*

According to the results given in the "Theory of Magnets" paper a material surface experiences a tension $(E^2 - H^2)/8\pi$ per unit area of surface. If $E^2 - H^2$ at any point of the surface is written in the form

$$A_0 + A_1\cos\xi + A_2\cos 2\xi + \dots,$$

the coefficients A_n are power-series in k, each starting with a term of higher order than the previous coefficient. In order to make the tension constant over the surface, it would be necessary, as in the corresponding hydrodynamical problem of a hollow vortex ring, to take into account the fact that the section of the ring cannot be a circle, and to find its equation in the form

$$\eta = \eta_0 + \beta_1 \cos \xi + \beta_2 \cos 2\xi + \ldots$$

under the condition of constant pressure. Since, however, the section is circular if terms in $\cos \xi$ are included, the necessary condition that the pressure should be constant over the surface is, to the order of approximation used throughout, that A_1 should vanish. The matter has then to stand a uniform hydrostatic pressure which can be calculated from the value of A_0.

The resultant force E at the surface is normal to it, and equal to

$$\frac{1}{J}\frac{\partial \phi}{\partial \eta} = \frac{1}{J}\sinh \eta \frac{\partial \phi}{\partial z} = \frac{z + \cos \xi}{a}\sinh \eta \frac{\partial \phi}{\partial z}. \quad \ldots\ldots(31)$$

Similarly the resultant magnetic force H is tangential to the surface, and equal to

$$\frac{1}{\rho J}\frac{\partial \psi}{\partial \eta} = \frac{1}{\rho J}\sinh \eta \frac{\partial \psi}{\partial z} = \frac{(z + \cos \xi)^2}{a^2}\frac{\partial \psi}{\partial z}. \quad \ldots\ldots(32)$$

The value of $\dfrac{\partial \phi}{\partial z}$ is given in equation (20), and that of $\dfrac{\partial \psi}{\partial z}$ in equation (26 a), which, on substituting for the coefficients b_n, can be written in the form

$$-\frac{1}{2}\frac{\psi_0}{(z + \cos \xi)} + (z + \cos \xi)^{-\frac{1}{2}}\frac{\psi_0}{\pi \sqrt{2}}\left\{\frac{P_0 \dfrac{dQ_0}{dz}}{\dfrac{dP_0}{dz}} + 2\sum_1^\infty (-)^n \frac{P_n \dfrac{dQ_n}{dz}}{\dfrac{dP_n}{dz}}\right\}$$

$$= \psi_0 (z + \cos \xi)^{-\frac{1}{2}}$$

$$\times \left[-\frac{1}{2}(z + \cos \xi)^{-\frac{1}{2}} + \frac{1}{\pi \sqrt{2}}\left\{\frac{P_0 \dfrac{dQ_0}{dz}}{\dfrac{dP_0}{dz}} + 2\sum_1^\infty (-)^n \cos n\xi \frac{P_n \dfrac{dQ_n}{dz}}{\dfrac{dP_n}{dz}}\right\}\right].$$

Since

$$(z + \cos \xi)^{-\frac{1}{2}}\frac{\pi}{\sqrt{2}} = Q_0 + 2\sum_1^\infty Q_n \cos n\xi,$$

the above result can be written in the form

$$\frac{\psi_0 (z + \cos \xi)^{-\frac{1}{2}}}{\pi \sqrt{2}}$$

$$\times \left\{ \left(\frac{P_0 \dfrac{dQ_0}{dz}}{\dfrac{dP_0}{dz}} - Q_0 \right) + 2 \Sigma (-)^n \cos n\xi \left(\frac{P_n \dfrac{dQ_n}{dz}}{\dfrac{dP_n}{dz}} - Q_n \right) \right\}$$

$$= - \frac{\psi_0 (z + \cos \xi)^{-\frac{1}{2}}}{\sqrt{2} \sinh^2 \eta} \left\{ \frac{1}{\left(\dfrac{dP_0}{dz} \right)} + 2 \Sigma (-)^n \cos n\xi \frac{1}{\left(\dfrac{dP_n}{dz} \right)} \right\}, \dots (33)$$

using results given by Basset, p. 27.

We can now calculate the value of $E^2 - H^2$, which is equal to

$$\frac{(z + \cos \xi)^3}{a^2 \sinh^2 \eta} \left\{ \frac{(\phi_0 - e/a)^2}{P_0^2} - 4 \cos \xi \frac{(\phi_0 - e/a)(e/a + \phi_0)}{P_0 \quad P_1} \right\}$$

$$- \frac{(z + \cos \xi)^3 \psi_0^2}{2a^4 \sinh^4 \eta} \left\{ \frac{1}{\left(\dfrac{dP_0}{dz} \right)^2} - \frac{4 \cos \xi}{\left(\dfrac{dP_0}{dz} \right) \left(\dfrac{dP_1}{dz} \right)} \right\},$$

where in the brackets only the constant term and the term in $\cos \xi$ have been retained.

Expanding $(z + \cos \xi)^3$ to the same order, $E^2 - H^2$ is finally equal to

$$\left\{ \frac{2z^3}{a^2 \sinh^2 \eta} \frac{(\phi_0 - e/a)^2}{P_0^2} - \frac{\psi_0^2 z^3}{2a^4 \sinh^4 \eta} \frac{1}{\left(\dfrac{dP_0}{dz} \right)^2} \right\}$$

$$+ \frac{2 \cos \xi}{a^2 \sinh^2 \eta} \left\{ \frac{3z^2}{P_0^2} (\phi_0 - e/a)^2 - \frac{4z^3 (\phi_0^2 - e^2/a^2)}{P_0 P_1} \right\}$$

$$- \frac{\psi_0^2 \cos \xi}{2a^4 \sinh^4 \eta} \left\{ \frac{3z^2}{\left(\dfrac{dP_0}{dz} \right)^2} - \frac{4z^3}{\left(\dfrac{dP_0}{dz} \right) \left(\dfrac{dP_1}{dz} \right)} \right\} \dots \dots \dots \dots (34)$$

Making the coefficient of $\cos \xi$ vanish we have, to the order of approximation required,

$$\frac{3}{L^2} (\phi_0 - e/a)^2 - \frac{2}{L} (\phi_0^2 - e^2/a^2) - \frac{\psi_0^2}{a^2} \left\{ \frac{3}{(L-2)^2} + \frac{2}{(L-2)} \right\} = 0,$$
$$\dots \dots (35)$$

or, replacing $\phi_0 - e/a$ by $Le_m/a\pi$, by (22),

$$(3 - 2L) e_m^2 - 4\pi e\, e_m - \pi^2 \psi_0^2 (2L - 1)/(L - 2)^2 = 0 \dots (36)$$

The constant value of $E^2 - H^2$ is, to the same order, then equal to

$$\frac{1}{4k^2a^4\pi^2}\left\{e_m{}^2 - \frac{\pi^2\psi_0{}^2}{(L-2)^2}\right\}. \quad \ldots\ldots\ldots\ldots(37)$$

If the central charge $e = - e_m$, so that the magneton plus electron is neutral, then (36) becomes

$$(3 + 4\pi - 2L)\, e_m{}^2 - \pi^2\psi_0{}^2\,(2L-1)/(L-2)^2 = 0.\ldots(38)$$

§ 5. *The electromagnetic energy of the magneton plus electron.*

The electric and magnetic energies can be calculated separately. As regards the former, it is equal to

$$\frac{1}{2}\int \phi\rho\,dx\,dy\,dz,$$

where ϕ is the electric potential, and ρ the volume density of electrification.

The contribution of the electron itself may be written E_c, which denotes the intrinsic energy of the central electron. For the mutual influence of the electron and magneton, the energy is

$$\tfrac{1}{2}e\phi_1,$$

where ϕ_1 is the potential of the magneton at the centre of the ring, which is e_m/a to the order of approximation involved, so that the mutual energy is $\tfrac{1}{2}ee_m/a$. As regards the contribution of the surface charge on the magneton, this is clearly

$$\frac{1}{2}\,\phi_0\,e_m = \frac{1}{2}\,e_m\left(\frac{e}{a} + \frac{Le_m}{a\pi}\right).$$

Hence the total electric energy is

$$E_c + \frac{1}{2}\,ee_m/a + \frac{1}{2}\,e_m\left(\frac{e}{a} + \frac{Le_m}{a\pi}\right),$$

or $$E_c + ee_m/a + \frac{1}{2}\frac{L}{a\pi}\,e_m{}^2. \quad \ldots\ldots\ldots\ldots(39)$$

As regards the magnetic energy, this is calculated as in the corresponding hydrodynamical problem (Basset, Vol. I, p. 85), giving

$$\frac{1}{8\pi}\int H^2\,dx\,dy\,dz$$

$$= \frac{1}{8\pi}\int 2\pi\rho\,d\rho\,dz\,\frac{1}{\rho^2}\left[\left(\frac{\partial\psi}{\partial\rho}\right)^2 + \left(\frac{\partial\psi}{\partial z}\right)^2\right]$$

$$= \frac{1}{4}\int\frac{\psi}{\rho}\left(\frac{\partial\psi}{\partial\rho}\,dz + \frac{\partial\psi}{\partial z}\,d\rho\right) - \frac{1}{4}\iint\frac{\psi}{\rho}\left(\frac{\partial^2\psi}{\partial\rho^2} + \frac{\partial^2\psi}{\partial z^2} - \frac{1}{\rho}\frac{\partial\psi}{\partial\rho}\right)d\rho\,dz.$$

The double integral vanishes owing to the differential equation satisfied by the stream function ψ, and the single integral can be expressed as an integral along the meridian curve of the boundary of the region.

Hence the magnetic energy is equal to

$$\frac{1}{4} \int \frac{\psi}{\rho} \frac{\partial \psi}{\partial n} \, ds = \frac{\psi_0}{4} \int \frac{1}{\rho} \frac{\partial \psi}{\partial \eta} \, d\xi$$

$= \frac{1}{4} \psi_0$ (circulation through the aperture of the ring),

so that, using the result of (29), the magnetic energy is

$$\frac{1}{4} \psi_0 \frac{2\pi \psi_0}{a(L-2)} \quad \text{or} \quad \frac{\pi \psi_0^2}{2a(L-2)}. \quad \dots\dots\dots(40)$$

The total electromagnetic energy is therefore

$$E_c + e\,e_m/a + \frac{1}{2} \frac{L e_m^2}{a\pi} + \frac{\pi \psi_0^2}{2a(L-2)}. \quad \dots\dots(41)$$

§ 6. *Small oscillations of the central electron in the field of the magneton.*

If the central electron is slightly disturbed from its position of equilibrium it will, if a certain condition is satisfied, oscillate about this position. If we were considering only the electro-static force due to the surface charge on the anchor ring, the equilibrium would be unstable for displacements in the plane of the ring and stable for displacements perpendicular to this plane, as in the corresponding gravitational case. If the magnetic force due to the magneton is included and is sufficiently large, the central position of the electron is stable for all displacements. This provides an interesting example of the well-known effect of gyrostatic forces in converting instability into stability. [Cf. e.g. Lamb, *Higher Mechanics*, p. 240.]

The electric and magnetic forces at points near the centre of the ring ($\xi = 0$, $\eta = 0$) can be calculated without difficulty, the components of the former being to a first approximation

$$-e_m \rho/2a^3, \quad +e_m z/a^3 \dots\dots\dots\dots(42)$$

in the directions perpendicular to and along the axis of z.

The components of the magnetic force in the same directions are to the same order

$$0, \quad \psi_0 \pi/a^2 (L-2). \quad \dots\dots\dots(43)$$

If m denotes the mass of the central electron, $-e$ its charge, the equations of motion for small oscillations are

$$
\left.
\begin{aligned}
m\ddot{x} + \frac{e}{c}\dot{y}H_0 - \frac{\dot{e}\,e_m}{2a^3}\,x &= 0 \\[2mm]
m\ddot{y} - \frac{e}{c}\dot{x}H_0 - \frac{e\,e_m}{2a^3}\,y &= 0 \\[2mm]
m\ddot{z} \qquad\;\; + \frac{e\,e_m}{a^3}\,z &= 0
\end{aligned}
\right\}, \quad\ldots\ldots\ldots\ldots(44)
$$

where H_0 stands for $\psi_0\,\pi/a^2\,(L-2)$.

The third equation gives the period for stable oscillations perpendicular to the plane of the ring, and the other two equations give the periods of oscillations in this plane. Putting $e e_m/2ma^3 = \lambda$, $eH_0/mc = \mu$, these latter equations become

$$\ddot{x} + \mu\dot{y} - \lambda x = 0, \quad \ddot{y} - \mu\dot{x} - \lambda y = 0,$$

or, if $z = x + iy$,

$$\ddot{z} - i\mu\dot{z} - \lambda z = 0. \quad\ldots\ldots\ldots\ldots\ldots(45)$$

This gives the solution $z = Ae^{ipt}$, where

$$p^2 - \mu p + \lambda = 0,$$

or $\qquad\qquad p = \tfrac{1}{2}\{\mu \pm \sqrt{(\mu^2 - 4\lambda)}\},$

giving a real period if $\mu^2 > 4\lambda$, or

$$H_0^2 > 2mc^2 e_m/ea^3. \quad\ldots\ldots\ldots\ldots\ldots(46)$$

If μ is very large in comparison with λ, i.e. the magnetic force completely overpowers the electric force, the first two of equations (44) become

$$m\ddot{x} - \frac{eH_0}{c}\dot{y} = 0, \quad m\ddot{y} + \frac{eH_0}{c}\dot{x} = 0, \quad\ldots\ldots\ldots(47)$$

giving a circular motion of period $2\pi/p$, where $p = eH_0/mc$.

§ 7. Numerical Magnitudes and Physical Applications.

Without some modification in the theory it does not seem possible to make much progress in its application to physical phenomena. Dealing with the case of a magneton whose positive charge is e_m, together with a central electron whose charge is e, the chief results of the preceding theory are:

(a) Angular Momentum $= e_m\psi_0/c$.

(b) Magnetic Moment of Magneton $= \dfrac{\pi a\,\psi_0}{2\,(L-2)}$. $\ldots\ldots\ldots(30)$

(c) Electromagnetic Energy

$$= E_c + \frac{e\,e_m}{a} + \frac{1}{2}\frac{L e_m{}^2}{a\pi} + \frac{\pi\psi_0{}^2}{2a\,(L-2)}. \quad\ldots\ldots\ldots(41)$$

(d) Surface Condition

$$(3 - 2L)\,e_m{}^2 - 4\pi e\,e_m - \pi^2\psi_0{}^2\,\frac{(2L-1)}{(L-2)^2} = 0.\ldots\ldots(36)$$

(e) Condition of stability of central electron

$$\frac{\psi_0{}^2\pi^2}{a^4\,(L-2)^2} > \frac{2mc^2 e_m}{ea^3}, \quad\ldots\ldots(43) \text{ and } (46)$$

or

$$a < \frac{\psi_0{}^2\pi^2 e}{(L-2)^2\,2mc^2 e_m}.\ldots\ldots\ldots\ldots(48)$$

Professor McLaren identifies the angular momentum with the value $h/2\pi$, where h is Planck's constant, so that

$$e_m\psi_0/c = h/2\pi, \quad\ldots\ldots\ldots\ldots\ldots(49)$$

and takes the magnetic moment I of the magneton to be given by Weiss' value $18\cdot54 \times 10^{-22}$ E.M.U.

Consider first of all the question of the existence of an isolated magneton on this theory. Putting $e = 0$, we have to determine L, a, e_m, ψ_0 from the equations (49), (30), (36) and (41), the two latter becoming in this case

$$(3 - 2L)\,e_m{}^2 = \pi^2\psi_0{}^2\,(2L-1)/(L-2)^2, \quad\ldots\ldots(50)$$

and   Electromagnetic Energy $= \dfrac{1}{2}\dfrac{L e_m{}^2}{a\pi} + \dfrac{\pi\psi_0{}^2}{2a\,(L-2)},$

or, using (50), $= \dfrac{\pi\psi_0{}^2}{2a\,(L-2)^2}\dfrac{6\,(L-1)}{(3-2L)}.\ldots(51)$

From these equations it is clear that L must lie between the values 1 and $\frac{3}{2}$, and, taking the larger value, this would give $k/4 = e^{-\frac{3}{2}}$ or k nearly equal to unity, which would not be in accordance with the assumptions under which the mathematical calculations were made in which k is assumed to be a small quantity.

In the case of a neutral system in which $e = -e_m$, equations (36) and (41) become

$$(3 + 4\pi - 2L)\,e_m{}^2 = \pi^2\psi_0{}^2\,(2L-1)/(L-2)^2, \quad\ldots(52)$$

and Electromagnetic Energy

$$= E_c + (L - 2\pi)\,e_m{}^2/2a\pi + \pi\psi_0{}^2/2a\,(L-2),$$

or, using (52),

$$= E_c + \frac{\pi\psi_0^2}{2a} \frac{6\,(L-1-\pi)}{(L-2)^2\,(3+4\pi-2L)}. \quad \ldots\ldots\ldots\ldots(53)$$

If this last equation is used to define the electromagnetic mass M of the magneton by the equation

Electromagnetic Energy = (mass) c^2,

then
$$Mc^2 = \frac{\pi\psi_0^2}{2a} \frac{6\,(L-1-\pi)}{(L-2)^2\,(3+4\pi-2L)}. \quad \ldots\ldots\ldots(54)$$

From (52) and (54) we should expect L to lie between $\frac{1}{2}(3+4\pi)$ and $1+\pi$, i.e. between 7·78 and 4·14, the latter of which would make k equal to about 0·07, which would satisfy the conditions of the mathematical analysis.

Eliminating ψ_0 from (49) and (52), we obtain the relation

$$\frac{(3+4\pi-2L)\,(L-2)^2}{(2L-1)} = \frac{h^2c^2}{4e_m^4} \quad \ldots\ldots\ldots\ldots(55)$$

Since for the range of possible values of L the maximum value of the left-hand side of this equation is about 5·7, the value e_m of the charge on the magneton would have to be at least 13 times as large as the charge on an electron in order to be able to satisfy this equation, so that the neutral system considered could not apply to the case of a hydrogen atom.

If e_m is taken as the charge on an electron, we then have four equations (49), (52), (54) and (30) from which to find the three quantities a, ψ_0 and L. Omitting any one of the four equations, e.g. the angular momentum equation (49), we might attempt to find the three unknown quantities from the values of M (the mass of the hydrogen nucleus), and I (the magnetic moment), using equations (52), (54) and (30), but it will be found impossible to find a value of L lying in the range of possible values. The same result is true if any one of the four equations be omitted, so that there would appear to be no way possible of applying the theory to the simple case of a hydrogen atom, and the application to any more complicated case would need further preliminary mathematical analysis in regard to the field of the electrons in the atom.

SECTION III. THE PROPAGATION OF AN ARBITRARY DISTURBANCE IN A DISPERSIVE MEDIUM

[EDITORIAL NOTE: BY PROF. T. H. HAVELOCK.

The third group of papers left by Professor McLaren deals with the subject of the propagation of a disturbance in a dispersive medium, and consists of a comparison between the mathematical analysis by Fourier integrals and the physical separation of an impulse into waves and groups of waves.

It has not been found possible to reproduce all the investigation as a complete paper owing to the existence of gaps in the manuscript. What follows consists of the first seven sections of the work, and deals with a certain type of medium for which exact solutions are given for some forms of initial impulses. It appears to have been Professor McLaren's intention to publish this as a separate paper.

The remaining portion of the papers, not reproduced here, deals with similar problems for a medium containing vibrating electrons with a single natural period. McLaren considers a limited impulse incident upon the face of such a medium. The electric and magnetic vectors in the reflected and transmitted disturbances are obtained in the form of contour integrals in the plane of a complex variable. Exact solutions are not obtained, but McLaren discusses in detail methods of approximation by suitable deformation of the path of integration. His analysis seems in fact to have been an independent study in the method now generally known as the "method of steepest descents," and first used in this connection by Brillouin (*Ann. der Phys.* Bd. 44, p. 203, 1914), who applied it to the approximate evaluation of the integrals obtained by Sommerfeld for the propagation of a limited train of simple harmonic waves.

The numbering of the equations has been left as it is in the manuscript; equations (1) and (2) are missing, together with the first page or two of the introduction to the paper.]

The Fourier Analysis of the Energy gives a result which depends only on the nature of the disturbance issuing from the source; but that Analysis is not in itself a physical reality. Actual dispersion gives results which depend on the nature of the substance used. The question arises how far the process which goes on in fact corresponds to the results indicated by the Analysis. Professor Larmor (*Phil. Mag.*, Vol. 10, p. 575, 1905) points out that the dispersion we actually observe in nature is for the most part due to the fact that the medium consists of molecules which have periods of vibration of their own. The Fourier Analysis then leads to the result that the energy of vibration may become infinite and this result requires interpretation. Even when the Analysis leads to a definite result it does not show the stages through which a disturbance initially given arbitrarily passes into the wave form. In this paper I show how a disturbance at first confined to a limited area spreads out into a regular wave train, and I prove that the results obtained by Fourier's method are in general adequate to represent the facts.

In § 1 it is shown without using the Fourier series that a prism of finite dispersing power does produce approximately simple harmonic waves. The proof does not explain the manner in which dispersion arises, but it does not involve any assumption as to the "regularity" of the incident light.

In the following sections I use the Electromagnetic theory of light in the form given to it by Lorentz. Certain simple cases of dispersion give results which are expressible in Bessel Functions. The curves given in Figs. 3 to 5 and 6 to 10 illustrate such cases. The difficulty raised by Larmor does not, it seems, indicate a failure of Fourier's method. In the absence of any true absorption there will be an actual accumulation of energy in the molecular vibrations which is correctly indicated by the Fourier Integral.

§ 1. *Dispersion by a prism of finite resolving power.*

The medium in which the dispersion takes place must be such that trains of simple harmonic waves can travel in it without change of type. Evidently this is necessary in order that dispersion may exist at all.

Let ϕ be the vector quantity, electromotive force for example, which defines the disturbance in a plane wave travelling along the axis of x. Then ϕ must satisfy a differential equation of the form

$$\Sigma a_{mn} \left(\frac{d}{dt}\right)^m \left(\frac{d}{dx}\right)^n \phi = 0. \dots\dots\dots\dots(3)$$

The line $\alpha\beta$ divides the medium on the left from æther on the right.

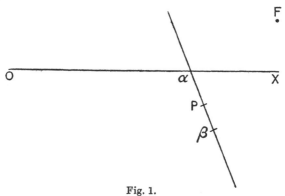

Fig. 1.

We are to find the disturbance at F due to the light which crosses between the points α and β. The disturbance at F at any time is due to contributions from points between α and β; these arrive simultaneously at F and therefore passed at different times across the line $\alpha\beta$. P is any point between α and β, and the projection of αP on OX is equal to x. Suppose $\alpha\beta$ small compared with αF. The time an element of the disturbance takes to travel from P to F will differ from the time taken from α to F by a quantity proportional to x. Let the difference be $\frac{x}{v}$. Then a disturbance starting at time t_1 from α arrives at F at the same time as another starting from P at time $t_1 + \frac{x}{v}$.

The total disturbance at F is therefore proportional to Φ, where we have

$$\Phi = \int_0^a \phi \left(t_1 + \frac{x}{v} \right) dx, \quad \dots\dots\dots\dots(4)$$

and a is the value of x at β.

Imagine a point to start from α at time t_1 and to travel along $\alpha\beta$ with a velocity v parallel to the axis OX. $\phi \left(t_1 + \frac{x}{v} \right)$ is evidently the value of ϕ in each position reached by the moving point. We may transform equation (3) by using the variables t_1 and x instead of t and x.

We have $\dfrac{d}{dt} = \dfrac{\partial}{\partial t_1}$, and $\dfrac{d}{dx} = \dfrac{\partial}{\partial x} - \dfrac{1}{v}\dfrac{\partial}{\partial t_1}$, so that (3) becomes

$$\Sigma a_{mn} \left(\frac{\partial}{\partial t_1} \right)^m \left(\frac{\partial}{\partial x} - \frac{1}{v}\frac{\partial}{\partial t_1} \right)^n \phi = 0, \quad \dots\dots\dots\dots(5)$$

which may be written

$$\Sigma (-1)^n \frac{a_{mn}}{v^n} \left(\frac{\partial}{\partial t_1} \right)^{m+n} \phi = \frac{\partial}{\partial x}(\phi_1). \quad \dots\dots\dots\dots(6)$$

In (6) all the terms of (5) which do not contain $\dfrac{\partial}{\partial x}$ are on the left-hand side, and the terms containing $\dfrac{\partial}{\partial x}$ are on the right and we may write $\dfrac{\partial \phi_1}{\partial x}$ for these.

(4) is now simply $\qquad \Phi = \int_0^a \phi \, dx. \quad \dots\dots\dots\dots\dots(7)$

(6) can be integrated with respect to x between the limits 0 and a. Then

$$\Sigma (-1)^n \frac{a_{mn}}{v^n} \left(\frac{\partial}{\partial t_1} \right)^{m+n} \Phi = [\phi_1]_0^a. \quad \dots\dots\dots\dots(8)$$

The right-hand side of (8) is the difference in the value of ϕ_1 between the limits 0 and a.

The only assumption we make as to the character of the undispersed light is that the average value of Φ as given by (7) increases as a is increased. Then by making a great we can make Φ on the average so large that the right-hand side of equation (8) may be neglected compared with the terms on the left. We then have

$$\Sigma \, (-1)^n \frac{a_{mn}}{v^n} \left(\frac{\partial}{\partial t_1}\right)^{m+n} \Phi = 0. \quad \ldots\ldots\ldots\ldots(9)$$

Any solution of (9) is of the form

$$\Phi = A \, e^{ipt},$$

and

$$\Sigma \, (-1)^n \frac{a_{mn}}{v^n} (ip)^{m+n} = 0,$$

or

$$\Sigma \, (-1)^n \, a_{mn} \left(\frac{-ip}{v}\right)^n (ip)^m = 0.$$

This last equation is the condition that (3) may have a solution of the form $e^{ip}\left(t - \frac{x}{v}\right)$. If therefore waves of period $\frac{2\pi}{p}$ travel in the medium with velocity v the disturbance at F due to dispersion is approximately simple harmonic in this period.

The colours obtained are evidently the purer the larger a is made but the proof does not show in detail how the dispersion arises. We now consider the Electromagnetic Theory of Dispersion.

§ 2. Imagine a medium containing n molecules in unit volume each having one electron whose period of free vibration is $\frac{2\pi}{p}$; E and H are the electric and magnetic forces in a plane wave which travels along the axis of x. E and H may be taken at right angles to each other. The mass of each electron, a mass certainly in part if not altogether electromagnetic in origin, is m, its charge is e.

The electromagnetic equations are

$$\frac{dE}{dt_1} + 4\pi n e \frac{dz_1}{dt_1} = - \, V \frac{dH}{dx_1}, \quad \ldots\ldots\ldots\ldots\ldots(10)$$

$$\frac{dH}{dt_1} = - \, V \frac{dE}{dx_1}, \quad \ldots\ldots\ldots\ldots\ldots(11)$$

$$m \frac{d^2 z_1}{dt_1^2} + m p^2 z_1 = eE + 4\pi n e^2 \left(\tfrac{1}{3} + \sigma\right) z_1. \quad \ldots\ldots(12)$$

z_1 is the displacement of an electron at time t_1 and the second term on the right-hand side of equation (12) represents the electric force due to the polarization of the medium, σ may probably be taken $= 0$ and in any case $\frac{1}{3} + \sigma$ will be taken positive.

The medium is to be treated in all that follows as if it were continuous. This makes it impossible to consider the dispersion of any disturbance which is not uniform over a distance large compared with the average distance between molecules. So far as the electrons are set in motion by the passage of such a disturbance the energy they receive will be scattered irregularly. This is the case for example with the Röntgen pulse. It is so thin that we have no regular reflection.

It may be thought that (12) ought to be

$$m \frac{d^2 z_1}{dt_1^2} + mp^2 z_1 - \frac{2}{3} \frac{e^2}{V^3} \frac{d^3 z_1}{dt_1^3} = eE + 4\pi n e^2 \left(\frac{1}{3} + \sigma \right) z_1.$$

Here the third term on the left represents the reaction on the electron of its own radiation. (See H. A. Lorentz, *Math. Ency.*, Band 5, p. 190). It plays a very important part in Planck's theory of dispersion but it is rightly omitted in (12). For in (12) E represents the average electromotive force in an element of volume containing a large number of molecules, it therefore includes a part due to the presence and motion of any one of the electrons we consider; E is not merely the force external to that electron.

Suppose dq is an element of electric charge at the point A forming part of an electron moving with velocity v. The electromotive force at P, where $AP = r$ and the angle between v and AP is θ, is in the direction of v:

$$dq \frac{\cos \theta}{r^2} - dq \frac{\dfrac{dv}{dt}(1 + \cos^2 \theta)}{V^2 \, r} + \frac{2}{3} \frac{dq}{V^3} \frac{d^2 v}{dt^2}. \quad \ldots\ldots(13)$$

In order to find the total force E_1 at P we must integrate with respect to dq including all electrons in the neighbourhood of the point P. The value of E_1 will evidently depend upon whether we consider a point P itself as belonging to one of the electrons or not. The first and second terms of the expression (13) contain r and are therefore greater near an electron, but the third term of (13) does not contain r at all; it is the same wherever the point P may be and so it is included in the average electromotive force at the point.

The equation of motion of an electron which has a mass m_0 not of electromagnetic origin is

$$m_0 \frac{d^2 z_1}{dt_1^2} + m p^2 z_1 = \int dq_1 E_1, \quad \dots\dots\dots\dots(14)$$

where dq_1 is an element of the charge of the electron.

The second term on the left of (14) includes any restoring force on the electron which is not supposed to arise from the electromagnetic energy.

E is the average electric force and referring to (13) it appears that E_1 differs from the average force E by two terms, one depending upon the displacement of the electrons and the other on their acceleration, or

$$\int dq_1 E_1 = eE + 4\pi n e^2 \left(\tfrac{1}{3} + \sigma\right) z_1 - m_1 \frac{d^2 z_1}{dt_1^2},$$

where m_1 is the electromagnetic mass, so that (14) becomes (12) when we write $m = m_0 + m_1$.

The term involving $\frac{d^3 z_1}{dt_1^3}$ if it existed would indicate a scattering of the light. It appears that this does not result so long as the medium may be treated as continuous.

It is convenient to change the scale of time and space in equations (10) to (12).

Write
$$t = t_1 \left(p^2 - \tfrac{4}{3}\pi \frac{ne^2}{m} - 4\pi \frac{ne^2\sigma}{m}\right)^{\frac{1}{2}},$$

$$x = \frac{x_1}{V} \left(p^2 - \tfrac{4}{3}\pi \frac{ne^2}{m} - 4\pi \frac{ne^2\sigma}{m}\right)^{\frac{1}{2}},$$

$$z = z_1 \frac{m}{e} \left(p^2 - \tfrac{4}{3}\pi \frac{ne^2}{m} - 4\pi \frac{ne^2\sigma}{m}\right)^{\frac{1}{2}},$$

$$\kappa^2 - 1 = \frac{4\pi \dfrac{ne^2}{m}}{p^2 - \tfrac{4}{3}\pi \dfrac{ne^2}{m} - 4\pi \dfrac{ne^2\sigma}{m}}.$$

Equations (10), (11), (12) become

$$\frac{dE}{dt} + (\kappa^2 - 1)\frac{dz}{dt} = -\frac{dH}{dx}, \qquad \ldots\ldots\ldots\ldots(15)$$

$$\frac{dH}{dt} = -\frac{dE}{dx}, \qquad \ldots\ldots\ldots\ldots\ldots(16)$$

$$\frac{d^2z}{dt^2} + z = E. \qquad \ldots\ldots\ldots\ldots\ldots(17)$$

The expression for the energy per unit volume I take to be

$$\tfrac{1}{2}(E^2 + H^2) + \tfrac{1}{2}(\kappa^2 - 1)\left\{\left(\frac{dz}{dt}\right)^2 + z^2\right\}.$$

The flux of this energy per unit time across any plane perpendicular to the axis is EH.

By eliminating E and H from these equations, we obtain

$$\frac{d^2}{dt^2}\left(\frac{d^2}{dt^2} + \kappa^2\right)z = \frac{d^2}{dx^2}\left(\frac{d^2}{dt^2} + 1\right)z, \ldots\ldots\ldots\ldots(18)$$

and by using (15) we get an equation of the same form for E,

$$\frac{d^2}{dt^2}\left(\frac{d^2}{dt^2} + \kappa^2\right)E = \frac{d^2}{dx^2}\left(\frac{d^2}{dt^2} + 1\right)E. \qquad \ldots\ldots\ldots(19)$$

A typical solution of (18) or (19) is $e^{i(\delta t + \lambda x)}$,

where
$$\lambda^2 = \delta^2\frac{(\delta^2 - \kappa^2)}{\delta^2 - 1} \ldots\ldots\ldots\ldots\ldots\ldots\ldots(20)$$

There is therefore a train of simple harmonic waves of period $\frac{2\pi}{\delta}$ if δ is greater than κ, or less than 1. If δ lies between 1 and κ, then λ is imaginary.

It is not possible to get exact numerical results for the dispersion of a pulse in the medium we have just considered. We may suppose that the electrons are free to move instead of being attached to the molecules. Instead of equation (12), we have simply

$$m\frac{d^2z_1}{dt_1^2} = eE. \qquad \ldots\ldots\ldots\ldots\ldots(21)$$

Elimination of H and z from (10), (11) and (21) gives

$$\frac{d^2E}{dt_1^2} + \frac{4\pi ne^2}{m}E = V^2\frac{d^2E}{dx_1^2}.$$

Write $\qquad t_0 = t_1 \left(\dfrac{4\pi n e^2}{m}\right)^{\frac{1}{2}}, \quad x = x_1 \left(\dfrac{4\pi n e^2}{m V^2}\right)^{\frac{1}{2}},$

$$4\pi n e z_1 = z, \text{ and } v = \dfrac{dz}{dt},$$

then

$$\dfrac{dE}{dt} + v = -\dfrac{dH}{dx}, \qquad \text{.....................(22)}$$

$$\dfrac{dH}{dt} = -\dfrac{dE}{dx}, \qquad \text{.....................(23)}$$

$$\dfrac{dv}{dt} = E, \qquad \text{.....................(24)}$$

$$\dfrac{d^2 E}{dt^2} + E = \dfrac{d^2 E}{dx^2}. \qquad \text{.....................(25)}$$

The equation for v is not of the same form as that for E. It is

$$\dfrac{d^2}{dt^2}\left(\dfrac{dv}{dt}\right) + \dfrac{dv}{dt} = \dfrac{d^2}{dx^2}\left(\dfrac{dv}{dt}\right). \qquad \text{..............(26)}$$

A typical solution of (25) is $e^{i(\delta t + \lambda x)}$, and

$$\lambda^2 = \delta^2 - 1. \qquad \text{.....................(27)}$$

§ 3. Equation (25) has been solved in the form of an integral involving Bessel Functions; the initial conditions when $x = 0$ or $t = 0$ being supposed given. (See Riemann-Weber, *Partielle Differential-Gleichungen.*) It is possible to obtain solutions expressed directly in terms of Bessel Functions for certain cases. We may first use equations (10), (11) and (21) to give a general explanation of the process of dispersion in a medium where the electrons are free; the explanation can be verified in the exact solutions given later.

By adding to each other and by subtracting from each other (10) and (11),

$$\dfrac{d}{dt_1}\left(\dfrac{E+H}{2}\right) + V\dfrac{d}{dx_1}\left(\dfrac{E+H}{2}\right) + 2\pi n e v_1 = 0, \quad \text{...(28)}$$

$$\dfrac{d}{dt_1}\left(\dfrac{E-H}{2}\right) - V\dfrac{d}{dx_1}\left(\dfrac{E-H}{2}\right) + 2\pi n e v_1 = 0, \quad \text{...(29)}$$

or

$$\dfrac{d}{d\tau_1}\left(\dfrac{E+H}{2}\right) + 2\pi n e v_1 = 0 \text{ and } \dfrac{d}{d\tau_2}\left(\dfrac{E-H}{2}\right) + 2\pi n e v_1 = 0,$$

$\frac{d}{d\tau_1}$ and $\frac{d}{d\tau_2}$ representing differentiation with respect to the time at points moving forward and moving backward with velocity V. When $v_1 = 0$, (28) and (29) show that $\frac{E+H}{2}$ remains constant at a point moving forward with velocity V, and $\frac{E-H}{2}$ at a point moving backward with the same velocity. We have in fact the well known result; there is a forward wave in which the electric and magnetic forces are each $\frac{E+H}{2}$, and a backward wave in which the electric force is $\frac{E-H}{2}$ and the magnetic force is $-\frac{E-H}{2}$. When v_1 is not zero, (28) and (29) show that these waves are modified by the motion of the electrons and dispersion results. In a time $d\tau_1$ the forces in the forward wave are increased by $-2\pi nev_1 d\tau_1$ or $-2\pi ne\,dx_1$, where dx_1 is the displacement of an electron in the time $d\tau_1$ and the backward wave is altered in the same way. In fact the displacement of the electrons causes secondary waves of electric and magnetic force.

Imagine a pulse which consists initially of a forward wave only, so that at first $E = H$ throughout the pulse. The medium ahead of the pulse is undisturbed, so that there is no backward wave at the front of the pulse or $E = H$ there always. The forward wave moves with velocity V and v_1, the velocity of the ions, is always zero at the front, so that $\frac{d}{d\tau_1}\left(\frac{E+H}{2}\right) = 0$; or $\frac{E+H}{2}$ is constant and $\frac{E-H}{2} = 0$ at that point. E and H therefore remain unaltered at the front of the pulse as it moves forward; in fact since the electrons are not in motion till the wave begins to pass them, they have no effect in modifying the wave front.

Consider a point which at time τ is a distance y behind the front of the advancing disturbance, and which therefore passed it at time $\tau - \frac{y}{V}$.

(21) gives $$ev_1 = \frac{e^2}{m} \int_{\tau - \frac{y}{V}}^{\tau} E dt_1. \qquad \dots\dots\dots\dots(30)$$

The integration with respect to the time in (30) is at a point fixed in space or which moves backwards with a velocity V relative to the front, starting thence at time $\tau - \frac{y}{V}$, and reaching the point y at time τ.

(28) gives $$\frac{E + H}{2} = E_0 - 2\pi n \int_0^\tau ev_1 d\tau_1. \qquad \dots\dots\dots\dots(31)$$

$d\tau_1$ indicates integration with respect to the time at a point moving forward with velocity V or fixed at a distance y behind the front. E_0 is the initial value of $\frac{E + H}{2}$, since initially $E = H$.

Similarly equation (29) gives

$$\frac{E - H}{2} = -2\pi n \int_{\tau - \frac{y}{2V}}^{\tau} ev_1 d\tau_2, \qquad \dots\dots\dots\dots(32)$$

$d\tau_2$ indicating integration at a point moving backward from the front with velocity $2V$ and reaching a distance y at time τ.

From (31) and (32),

$$E = E_0 - 2\pi n \int_0^\tau ev_1 d\tau_1 - 2\pi n \int_{\tau - \frac{y}{2V}}^{\tau} ev_1 d\tau_2. \qquad \dots(33)$$

Suppose the pulse is of small thickness, and that the electric force within it is positive. The equations (30) and (33) indicate how dispersion arises. (30) shows that ev_1 is at first everywhere positive.

It follows from (33) that E decreases within the pulse, and is negative behind it. Hence from (30), ev_1 will ultimately become negative at some distance behind the pulse. If we return to (33), the integrals now contain elements for which ev_1 is negative, and E will in time become positive. In this way, as we shall see, regions in which E is positive alternate with those in which it is negative.

§ 4. I consider first the reflection and dispersion of a disturbance incident normally on the plane $x = 0$, which separates æther from the medium represented by the equations (22) to (25). E, the electric force, satisfies in æther the equation

$$\frac{d^2 E}{dt^2} = \frac{d^2 E}{dx^2},$$

since the velocity of light is 1 in the units we use.

A typical Fourier solution for a wave moving forward in æther is $E = e^{i(t-x)}$. The conditions at $x = 0$ are that E and $\dfrac{dE}{dx}$ are continuous across the plane $x = 0$. This gives easily

$$E = \frac{\delta - \lambda}{\delta + \lambda} e^{i\delta(t+x)}$$

for the reflected wave, and $E = \dfrac{2\delta}{\delta + \lambda} e^{i(\delta t - \lambda x)}$ for the transmitted wave. The transmitted wave moves forward from the plane $x = 0$. When λ is real, it is therefore to be of the same sign as δ; when λ is imaginary, $i\lambda$ is to be positive, so that the wave vanishes for large values of x. Suppose that E in the incident wave is equal to $f(t)$ when $x = 0$.

The ordinary Fourier analysis gives us for the transmitted wave

$$E = \frac{1}{\pi} \int_{-\infty}^{\infty} f(t_0)\, dt_0 \int_{-\infty}^{\infty} e^{i[\delta(t-t_0)-\lambda x]} \frac{\delta\, d\delta}{(\delta + \lambda)}, \quad \ldots(34)$$

$$H = \frac{1}{\pi} \int_{-\infty}^{\infty} f(t_0)\, dt_0 \int_{-\infty}^{\infty} e^{i[\delta(t-t_0)-\lambda x]} \frac{\lambda\, d\delta}{(\delta + \lambda)}, \quad \ldots(35)$$

$$v = \frac{1}{\pi} \int_{-\infty}^{\infty} f(t_0)\, dt_0 \int_{-\infty}^{\infty} e^{i[\delta(t-t_0)-\lambda x]} \frac{d\delta}{i(\delta + \lambda)}. \ldots(36)$$

An exact solution in terms of Bessel Functions can be got for the case where $f(t_0) = 0$ where t_0 is negative, and $f(t_0) = 1$ when t_0 is positive.

The case of a pulse of finite breadth in which the electric force is constant can be derived from this.

By (34) $\qquad E = \dfrac{1}{\pi} \displaystyle\int_0^\infty e^{-i\delta t_0}\, dt_0 \int_{-\infty}^\infty e^{i(\delta t - \lambda x)} \dfrac{\delta\, d\delta}{(\delta + \lambda)},$

or $\qquad E = \dfrac{1}{\pi} \displaystyle\int_{-\infty}^\infty e^{i(\delta t - \lambda x)} \dfrac{d\delta}{i(\delta + \lambda)}. \quad \ldots\ldots\ldots\ldots\ldots(37)$

The equation (27) is satisfied by writing

$$\delta = \frac{1}{2}\left(u + \frac{1}{u}\right) \text{ and } \lambda = \frac{1}{2}\left(u - \frac{1}{u}\right).$$

The path of integration for u is that indicated in Figure 2.

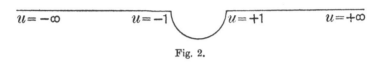

$$u = -\infty \qquad\qquad u = -1 \qquad\qquad u = +1 \qquad\qquad u = +\infty$$

Fig. 2.

It may easily be verified that the conditions as to λ are satisfied, and (37) becomes

$$E = \frac{1}{2\pi i} \int e^{\left[\frac{iu}{2}(t-x) + \frac{i}{2u}(t+x)\right]} \left(\frac{1}{u} - \frac{1}{u^3}\right) du. \quad \ldots\ldots(38)$$

On the circle $u = e^{-i\phi}$, where ϕ is between 0 and π,

$$\delta = \cos\phi, \text{ and } \lambda = -i\sin\phi \text{ or } i\lambda = \sin\phi;$$

so that $i\lambda$ is positive, as we required.

If $(t-x)$ is positive, $e^{\frac{iu}{2}(t-x)}$ vanishes at $u = \infty$ on the upper side of the real u axis; and if $(t-x)$ is negative, $e^{\frac{iu}{2}(t-x)}$ vanishes at $u = \infty$ on the lower side of the real u axis. The integral of (38) may therefore be taken round a closed contour by including the infinite half circle on the lower side of the real axis when $t-x$ is negative. This contour does not include any pole, and therefore E vanishes when $t-x$ is negative.

When $t-x$ is positive, the contour is completed by taking it round the infinite half circle on the upper side of the real axis. The complete contour then includes $u = 0$.

By writing $u = w\dfrac{(t+x)^{\frac{1}{2}}}{(t-x)^{\frac{1}{2}}}$, in (38), we have

$$E = J_0\left(t^2 - x^2\right)^{\frac{1}{2}} + \frac{(t-x)}{(t+x)} J_2\left(t^2 - x^2\right)^{\frac{1}{2}},$$

or, using the recurrence formula,

$$J_0(z) + J_2(z) = \frac{2}{z} J_1(z),$$

$$E = \frac{2x}{(t+x)} J_0\left(t^2 - x^2\right)^{\frac{1}{2}} + 2\frac{(t-x)^{\frac{1}{2}}}{(t+x)^{\frac{3}{2}}} J_1\left(t^2 - x^2\right)^{\frac{1}{2}} \ldots.(39)$$

When t is large, $(t^2 - x^2)^{\frac{1}{2}}$ is also large, except when $t - x$ is small. The first term on the right-hand side of (39) becomes large compared with the second. We have approximately

$$E = \left(\frac{2}{\pi}\right)^{\frac{1}{2}} \frac{2x}{(t + x)(t^2 - x^2)^{\frac{1}{4}}} \cos\left[(t^2 - x^2)^{\frac{1}{2}} - \frac{\pi}{4}\right]....(40)$$

(See Whittaker's *Modern Analysis*, p. 294.)

When x and t are large, put $x = \rho t$; a finite change in x or t produces only an infinitesimal change in ρ. (38) is equivalent to

$$E = \left(\frac{2}{\pi t}\right)^{\frac{1}{2}} \frac{2\rho}{(1 + \rho)^{\frac{3}{4}}(1 - \rho)^{\frac{1}{4}}} \cos\left[t(1 - \rho^2)^{\frac{1}{2}} - \frac{\pi}{4}\right]....(41)$$

A wave-length near the point x at time t corresponds to a change 2π in the quantity $t(1 - \rho^2)^{\frac{1}{2}}$. If $d\rho$ is the corresponding change in ρ, then $\dfrac{t\rho\, d\rho}{(1 - \rho^2)^{\frac{1}{2}}} = 2\pi$. The wave-length is $td\rho$, or

$$2\pi \frac{(1 - \rho^2)^{\frac{1}{2}}}{\rho} = 2\pi \frac{(t^2 - x^2)^{\frac{1}{2}}}{x}.$$

The short waves appear in front, where $\rho = 1$ or $x = t$; the long waves in the rear, where ρ is small. The velocity of the group, in which the wave-length is $2\pi\dfrac{(1 - \rho^2)^{\frac{1}{2}}}{\rho}$, is equal to ρ. The velocity of the waves themselves can be deduced from (40).

A crest moves so that $(t^2 - x^2)$ is a constant. This gives the velocity of the crest equal to $\dfrac{t}{x}$ or equal to $\dfrac{1}{\rho}$; ρ is always less than 1, so that the velocity of the waves is always greater than the velocity of the group.

The motion of the electrons, so far as that is oscillatory in character, might be deduced from (40) by integrating the equation $\dfrac{dv}{dt} = E$, with respect to the time, treating the harmonic term only as a variable. This does not however give that part of v which does not depend upon the time, and it is better to deduce the values of H and v directly from (35) and (36). By integrating with respect to t_0,

$$H = \frac{1}{\pi}\int_{-\infty}^{\infty} e^{i(\delta t - \lambda x)} \frac{\lambda d\delta}{i\delta(\delta + \lambda)}, \quad............(42)$$

$$v = -\frac{1}{\pi}\int_{-\infty}^{\infty} e^{i(\delta t - \lambda x)} \frac{d\delta}{\delta(\delta + \lambda)}. \quad.........(43)$$

The integrands in (42) and (43) have a pole at the point $\delta = 0$. In the medium which we are now considering, where the electrons are free, it is possible to have a steady current accompanied by magnetic, without electric, force. v and H are then functions of x, but not of t. The natural period of oscillation is therefore infinite; it is given by $\delta = 0$. The disturbance, whose reflection we now consider, is one in which $E = 0$ when t is negative, and $E = 1$ when t is positive.

The incident wave is in fact $\frac{1}{2}\left(e^{i\delta t} + e^{-i\delta t}\right)$ where $\delta = 0$, so that its period is the natural period of oscillation of the medium. We shall find that there is here no difficulty in consequence.

In (42) and (43) the path of integration, instead of keeping to the real axis, must be supposed to make a small half circle at the point $\delta = 0$ on the negative side. In Figure 2 the path must pass just below the point $u = -i$, instead of through it. Then H and v vanish when $t - x$ is negative, because the contour closed by the infinite half circle on the negative side does not include $\delta = 0$ or $u = -i$.

$$H = \frac{1}{2\pi i}\int e^{\frac{i}{2}(t^2 - x^2)^{\frac{1}{2}}\left(w + \frac{1}{w}\right)} \times \frac{\left[w^2\dfrac{(t+x)}{t-x} - 1\right]^2}{\left[w^2\dfrac{(t+x)}{t-x} + 1\right]} \frac{dw}{w^3}\frac{(t-x)}{(t+x)} \quad \ldots (44)$$

$$v = -\frac{1}{\pi}\int e^{\frac{i}{2}(t^2 - x^2)^{\frac{1}{2}}\left(w + \frac{1}{w}\right)} \times \frac{\left[w^2\dfrac{(t+x)}{(t-x)} - 1\right]}{\left[w^2\dfrac{(t+x)}{(t-x)} + 1\right]} \frac{dw}{w^2}\frac{(t-x)^{\frac{1}{2}}}{(t+x)^{\frac{1}{2}}} \quad \ldots (45)$$

In (44) and (45) the integrals are taken round a closed contour, including the origin and including the points

$$w = \pm i\frac{(t-x)^{\frac{1}{2}}}{(t+x)^{\frac{1}{2}}}.$$

When t and x are large the values of H and v, given by (44) and (45), may be deduced by the same method as that which gives the asymptotic values of the Bessel Functions. The contour in (44) and (45) must first be deformed so as to leave out the point $w = -i\frac{(t-x)^{\frac{1}{2}}}{(t+x)^{\frac{1}{2}}}$. The residue of H at this point is $2e^{-x}$.

The residue of v is also $2e^{-x}$. These quantities have to be added to the results obtained by taking the integrals of (44) and (45) round a contour not including $w = -i\dfrac{(t-x)^{\frac{1}{2}}}{(t+x)^{\frac{1}{2}}}$, but including $w = 0$ and $w = +i\dfrac{(t-x)^{\frac{1}{2}}}{(t+x)^{\frac{1}{2}}}$. This contour can be made up of the curve for which the real part of $w + \dfrac{1}{w}$ is 2 and the curve for which the real part of $w + \dfrac{1}{w}$ is -2. It will be found that the imaginary part of $w + \dfrac{1}{w}$ is everywhere positive along this contour. When $(t^2 - x^2)$ is great, $e^{\frac{i}{2}\left(w+\frac{1}{w}\right)(t^2-x^2)^{\frac{1}{2}}}$ is very small, except when $w = 1$ or $w = -1$.

Finally,

$$H = 2e^{-x} + \left(\frac{2}{\pi}\right)^{\frac{1}{2}} \frac{2x^2}{t(t+x)} \frac{\cos\left[(t^2-x^2)^{\frac{1}{2}} - \dfrac{\pi}{4}\right]}{(t^2-x^2)^{\frac{1}{4}}}, \quad \ldots\ldots(46)$$

$$v = 2e^{-x} + \left(\frac{2}{\pi}\right)^{\frac{1}{2}} \frac{2x(t-x)^{\frac{1}{2}}}{t(t+x)^{\frac{1}{2}}} \frac{\sin\left[(t^2-x^2)^{\frac{1}{2}} - \dfrac{\pi}{4}\right]}{(t^2-x^2)^{\frac{1}{4}}}. \quad \ldots(47)$$

It may be shown that the equations (40), (46) and (47) give a distribution of energy which agrees with the Fourier analysis of the energy entering the medium at the plane $x = 0$. The equations (37) and (42) do not give Fourier series expressing E and H as functions of x, for the range of λ includes imaginary as well as real quantities. (37) is a Fourier series expressing E as a function of t, but in (42) δ takes complex values over the small circle at $\delta = 0$.

Put $x = 0$ in (37),

$$E = \frac{1}{\pi} \int_{-\infty}^{\infty} \frac{e^{i\delta t} d\delta}{i(\delta + \lambda)}.$$

This gives a Fourier Integral, expressing E as a function of t valid for all values of t. It is therefore the only integral for t, and it follows that

$$\int_{-\infty}^{\infty} E e^{-i\delta t} dt = \frac{2}{i(\delta + \lambda)}, \quad \ldots\ldots\ldots\ldots(48)$$

which is valid for all real values of δ.

Write $x = 0$ in (42) also and break up the integral into five parts, the first from $\delta = -\infty$ to $\delta = -1$, the second from $\delta = -1$ to $\delta = -\epsilon$, ϵ being a small positive quantity. The third part is from $-\epsilon$ to $+\epsilon$ round the half circle, the fourth and fifth from ϵ to 1 and from 1 to ∞. When $\delta = 0$, $2\lambda = 1$, the integral of (42) taken round the half circle gives e^{-x}, λ when real is of the same sign as δ, when λ is imaginary, $i\lambda$ is positive, so that

$$H = 1 + \frac{1}{\pi} \int_{-\infty}^{-1} \frac{e^{i\delta t} \lambda d\delta}{i\delta (\delta + \lambda)} + \frac{1}{\pi} \int_{-1}^{-\epsilon} \frac{e^{i\delta t} \lambda d\delta}{i\delta (\delta + \lambda)} + \int_{\epsilon}^{1} \frac{e^{i\delta t} \lambda d\delta}{i\delta (\delta + \lambda)}$$
$$+ \int_{1}^{\infty} \frac{e^{i\delta t} \lambda d\delta}{i\delta (\delta + \lambda)} . \quad \ldots(49)$$

We can calculate $\int_{-\infty}^{\infty} EH dt$ from (46) and (47).

When $\delta = 0$, (46) gives $\int_{-\infty}^{\infty} E dt = 2$,

$$\int_{-\infty}^{\infty} EH dt = 2 + \frac{2}{\pi} \int_{-\infty}^{-1} \frac{\lambda d\delta}{\delta (\delta + \lambda)^2} + \frac{2}{\pi} \int_{1}^{\infty} \frac{\lambda d\delta}{\delta (\delta + \lambda)^2}$$
$$+ \frac{2}{\pi} \int_{-1}^{-\epsilon} \frac{\lambda d\delta}{\delta (\delta + \lambda) (\delta - \lambda)} + \frac{2}{\pi} \int_{\epsilon}^{1} \frac{\lambda d\delta}{\delta (\delta + \lambda) (\delta - \lambda)} . \quad \ldots(50)$$

In the first and second integrals of (50), λ is real and changes sign with δ. In the third and fourth, $i\lambda$ is positive. Remembering $\delta^2 - \lambda^2 = 1$, these terms are seen to cancel each other.

Finally $\qquad \int_{-\infty}^{\infty} EH dt = 2 + \frac{4}{\pi} \int_{1}^{\infty} \frac{\lambda d\delta}{\delta (\delta + \lambda)^2},$

or $\qquad \int_{-\infty}^{\infty} EH dt = 2 + \frac{4}{\pi} \int_{0}^{\infty} \frac{\lambda^2 d\lambda}{\delta^2 (\delta + \lambda)^2}. \quad \ldots\ldots\ldots(51)$

The distribution of energy indicated by (51) can be deduced from the equations (40), (46) and (47). We have seen that according to these equations, the waves at x at time t are nearly simple harmonic, and that the wave-length is $\dfrac{2\pi (t^2 - x^2)^{\frac{1}{2}}}{x}$.

Hence $\qquad \lambda^2 = \dfrac{x^2}{t^2 - x^2} = \dfrac{t^2}{t^2 - x^2} - 1,$

$$\frac{\lambda d\lambda}{dx} = \frac{t^2 x}{(t^2 - x^2)^2}. \quad \ldots\ldots\ldots\ldots\ldots(52)$$

By (40), (46) and (47), the total energy between x and $x + dx$ is on the average

$$= \frac{2}{\pi} \frac{x^2}{(t+x)^2} \frac{dx}{(t^2 - x^2)^{\frac{1}{2}}} + \frac{2}{\pi} \frac{x^4}{t^2 (t+x)^2} \frac{dx}{(t^2 - x^2)^{\frac{1}{2}}}$$

$$+ \frac{2}{\pi} \frac{x^2 (t-x)}{t^2 (t+x)} \frac{dx}{(t^2 - x^2)^{\frac{1}{2}}} + 4e^{-2x} dx$$

$$= \frac{2}{\pi} \frac{x^2 t^2 + x^4 + x^2 (t^2 - x^2)}{t^2 (t+x)^2 (t^2 - x^2)^{\frac{1}{2}}} dx + 4e^{-2x} dx$$

$$= \frac{4}{\pi} \frac{x^2 dx}{(t+x)^2 (t^2 - x^2)^{\frac{1}{2}}} + 4e^{-2x} dx. \quad \ldots\ldots\ldots\ldots\ldots(53)$$

Now $\qquad \lambda = \dfrac{x}{(t^2 - x^2)^{\frac{1}{2}}}$ and $\delta = \dfrac{t}{(t^2 - x^2)^{\frac{1}{2}}}$,

whence (52) gives

$$\frac{\lambda^2}{\delta^2} \frac{d\lambda}{(\delta + \lambda)^2} = \frac{x^2 dx}{(t+x)^2 (t^2 - x^2)^{\frac{1}{2}}}. \quad \ldots\ldots\ldots(54)$$

The second term of (53) gives, on integrating with respect to x, a total amount of energy equal to 2.

(54) shows that the distribution of energy given by (51) agrees with that given by (53). The Fourier analysis is therefore valid.

§ 5. Consider now a pulse of finite breadth in which the electric force is constant and equal to unity. Let this pulse be travelling in the æther, and let its front be incident upon the plane $x = 0$ at time $t = 0$, and let its rear reach the same plane at time $t = \tau$.

Then in the medium $[E = 0$ for $x > t]$

$$E = J_0 (t^2 - x^2)^{\frac{1}{2}} + \frac{t - x}{t + x} J_2 (t^2 - x^2)^{\frac{1}{2}}, \text{ if } \tau > t > 0, \, x < t, \ldots(55)$$

$[$and if $t > \tau, \, \tau < x < t]$

$$E = J_0 (t^2 - x^2)^{\frac{1}{2}} + \frac{t - x}{t + x} J_2 (t^2 - x^2)^{\frac{1}{2}} - J_0 [(t - \tau)^2 - x^2]^{\frac{1}{2}}$$

$$- \frac{t - \tau - x}{t - \tau + x} J_2 [(t - \tau)^2 - x^2]^{\frac{1}{2}}, \, t > \tau, \, 0 < x < \tau. \ldots(56)$$

The solution given by (56) is got by superimposing on the disturbance we considered in § 4, a second in which the electric force is -1 and begins at time $t = \tau$.

When $x = 0$ and $t > \tau$,

$$E = \frac{2}{t} J_1(t) - \frac{2}{(t-\tau)} J_1(t - \tau). \quad\ldots\ldots\ldots(57)$$

This equation shows that when $x = 0$, E becomes small when t is large; that is, compared with the period of the longest waves which are transmitted by the medium. The reflected wave therefore practically vanishes after that time.

The smaller the breadth of the pulse, the greater is the distance it can travel without being dispersed. At time t the front has travelled a distance t into the medium. The value of E given by (55) will be practically equal to 1, between $x = t$ and $x = t - \tau$, so long as $t^2 - (t - \tau)^2$ is small. The energy of the original pulse remains within its original breadth τ so long as $t\tau$ is small.

The result can be expressed in a form independent of the particular units we have used. If d is the distance the pulse has travelled, b the breadth of the pulse, and l the wave-length of the longest waves transmitted by the medium, the pulse is practically undispersed so long as $\dfrac{db}{l^2}$ is small.

When τ is small, we can find an approximate result for the energy reflected.

In the reflected wave

$$E = \frac{2}{t} J_1(t) - 1 \qquad\qquad (\tau > t), \;\ldots\ldots(58)$$

$$E = \frac{2}{t} J_1(t) - \frac{2}{t-\tau} J_1(t-\tau) \qquad (t > \tau). \;\ldots\ldots(59)$$

Also H is equal and opposite in value to E. The total energy reflected is $\displaystyle\int_0^\infty E^2 dt$, E being given by (58) or (59).

Now when t is large compared with τ, say of the order $\tau^{\frac{1}{2}}$ at least, (59) may be replaced by $2 \dfrac{d}{dt}\left(\dfrac{J_1(t)}{t}\right) \times \tau$.

When t is less than this, the expressions in (58) and (59) can be expanded in powers of t. The lowest power of t in $\dfrac{J_1(t)}{t}$ is t^2. Hence in (59) we shall have the lowest power of t derived from

$t^2 - (t - \tau)^2$. The greatest term in (58) or (59) is therefore one involving $t\tau$. The part of the energy reflected up to the time when $t = \sqrt{\tau}$ is proportional to $\int_0^{\sqrt{\tau}} t^2 \tau^2 dt$ or to $\tau^{\frac{7}{2}}$. This we shall find is very small compared with the total reflected energy.

Hence

$$\int_0^{\infty} E^2 dt = 4 \int_{\sqrt{\tau}}^{\infty} \left[\frac{d}{dt} \frac{J_1(t)}{t} \right]^2 dt \times \tau^2 \text{ approximately.}$$

When t is small, $\dfrac{d}{dt} \dfrac{J_1(t)}{t}$ is proportional to t, and if we again neglect a term of the order $\tau^{\frac{7}{2}}$, we have

$$\int_0^{\infty} E^2 dt = 4 \int_0^{\infty} \left[\frac{d}{dt} \frac{J_1(t)}{t} \right]^2 dt \times \tau^2$$

$$= 4 \int_0^{\infty} \left[\frac{J_2(t)}{t} \right]^2 dt \times \tau^2$$

$$= \frac{16}{3\pi} \tau^2.$$

(Shafheitlein, *Math. Ann.*, Bd 30, p. 168.)

The total energy of the pulse is τ. The fraction of this energy reflected is $\dfrac{16}{3\pi} \times \tau$.

In the case of the Röntgen pulse, the breadth estimated by experiments made on its diffraction is 10^{-8} cm. Suppose such a pulse incident on a metal surface. Metals cease to reflect perfectly when the wave-length of the incident light is less than that of the violet rays, say equal to 10^{-5} cms. τ is therefore at the most of the order 10^{-3} cms. And it appears that so far as the energy of the Röntgen pulse is regularly reflected, we should observe a reflecting power of the order $\dfrac{1}{10^3}$. In fact, the greater part of the reflection is due to scattering.

The breadth of the pulse is of the same order of magnitude as the average distance between the electrons, so that the secondary radiation due to the motion of any one electron as the pulse passes over it cannot interfere regularly with the radiation sent out by any other.

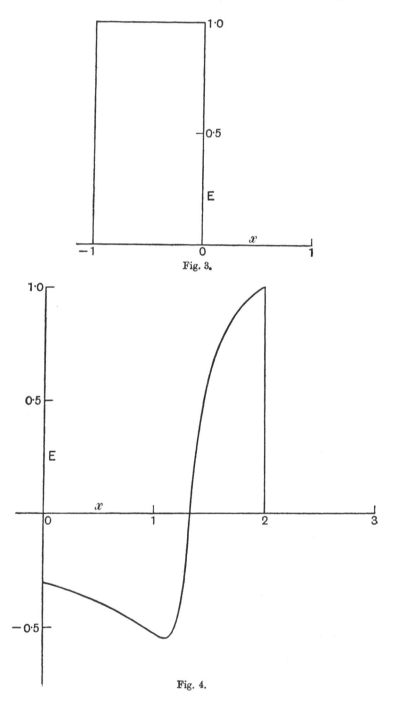

Fig. 3.

Fig. 4.

The figures 3, 4, 5 represent the electric force in the case of a pulse of unit breadth; Fig. 3 represents the pulse before it enters the medium, in Fig. 4 the electric force is represented after the front has travelled twice the breadth of the pulse, and in Fig. 5 after it has travelled four times the breadth. The first wave is seen in Fig. 5 emerging in the rear of the pulse.

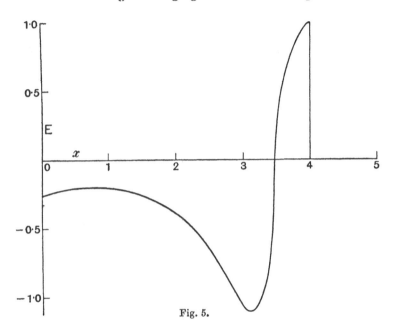

Fig. 5.

§ 6. Instead of a pulse incident on the medium from without, we may have a pulse originating within the medium and dispersed as it spreads outwards.

We may for example suppose that E is given at time $t = 0$ as a function of x, and that H and v are initially zero. Let $E = f(x_0)$ at time $t = 0$.

Then

$$E = \frac{1}{2\pi} \int_{-\infty}^{\infty} \frac{e^{i\delta t} + e^{-i\delta t}}{2} e^{i\lambda x} d\lambda \int_{-\infty}^{\infty} f(x_0) e^{-i\lambda x_0} dx_0. \quad \ldots(60)$$

Fourier's Theorem gives $E = f(x_0)$ when $t = 0$. Also (60) shows that $\dfrac{dE}{dt} = 0$ when $t = 0$, as ought to be the case according to (22), since v and H are initially zero. In (60) λ takes all real values, and δ is always real since $\delta^2 = \lambda^2 + 1$.

An interesting case, soluble in terms of Bessel Functions, is when

$$f(x_0) = f(-x_0); \quad \dots\dots\dots\dots\dots(61)$$

and for positive values of x_0,

$$f(x_0) = \int_1^\infty e^{-x_0 \alpha} \frac{d\alpha}{(\alpha^2 - 1)^{\frac{1}{2}}}. \quad \dots\dots\dots(62)$$

The integral with respect to α is a Bessel Function in which the argument is imaginary. When x_0 is small $f(x_0)$ is $-\log x_0$, and when x_0 is large, $f(x_0) = \left(\frac{\pi}{2x_0}\right)^{\frac{1}{2}} e^{-x}$.

The function $f(x_0)$ as defined by (62) has been tabulated by Isherwood (*Mem. and Proc. Manchester Lit. and Phil. Soc.* 1903–1904), who denotes the integral on the right of (62) by $K_0(x_0)$. But following the notation of H. Weber, I write

$$K_0(z) = \frac{2}{\pi} \int_1^\infty \cos z\alpha \frac{d\alpha}{(\alpha^2 - 1)^{\frac{1}{2}}}. \quad \dots\dots\dots(63)$$

(63) defines $K_0(z)$ for all real values of z, and $K_0(z)$ is the Bessel Function connected with Hankel's Function $Y_0(z)$ by the relation

$$K_0 = \frac{2}{\pi}\{-Y_0 + (\log 2 - \gamma) J_0\},$$

where γ is Euler's constant.

We then have

$$\int_{-\infty}^\infty f(x_0) e^{-i\lambda x_0} dx_0 = 2\int_0^\infty f(x_0) \cos \lambda x_0 dx_0$$

$$= 2\int_1^\infty \int_0^\infty e^{-x_0\alpha} \cos \lambda x_0 \frac{d\alpha}{(\alpha^2-1)^{\frac{1}{2}}} dx_0$$

$$= 2\int_1^\infty \frac{\alpha d\alpha}{(\alpha^2+\lambda^2)(\alpha^2-1)^{\frac{1}{2}}}$$

$$= 2\int_0^\infty \frac{d\beta}{\beta^2+1+\lambda^2} = \frac{\pi}{(1+\lambda^2)^{\frac{1}{2}}}. \quad \dots(64)$$

In (64), $(1+\lambda^2)^{\frac{1}{2}}$ is to be positive for all real values of λ. (60) now becomes

$$E = \int_0^\infty \cos \delta t \cos \lambda x \frac{d\lambda}{(1+\lambda^2)^{\frac{1}{2}}},$$

the integration being confined to positive values of λ. If we transform as before to the u plane,

$$E = \int_1^\infty \cos\frac{t}{2}\left(u + \frac{1}{u}\right)\cos\frac{x}{2}\left(u - \frac{1}{u}\right)\frac{du}{u}$$

$$= \frac{1}{2}\int_1^\infty \cos\left[\frac{1}{2}u\left(t+x\right) + \frac{1}{2u}\left(t-x\right)\right]\frac{du}{u}$$

$$+ \frac{1}{2}\int_1^\infty \cos\left[\frac{1}{2}u\left(t-x\right) + \frac{1}{2u}\left(t+x\right)\right]\frac{du}{u}$$

$$= \frac{1}{2}\int_0^\infty \cos\left[\frac{1}{2}u\left(t+x\right) + \frac{1}{2u}\left(t-x\right)\right]\frac{du}{u}. \quad\ldots\ldots\ldots\ldots(65)$$

(65) takes different forms, according as t is greater or less than x.

When $x > t$, write $u = w\,\dfrac{(x-t)^{\frac{1}{2}}}{(x+t)^{\frac{1}{2}}}$ in (65).

Then

$$E = \frac{1}{2}\int_0^\infty \cos\left[\tfrac{1}{2}\left(x^2 - t^2\right)^{\frac{1}{2}}\left(w - \frac{1}{w}\right)\right]\frac{dw}{w},$$

or $E = \dfrac{1}{4}\displaystyle\int_0^\infty e^{\frac{i}{2}(x^2-t^2)^{\frac{1}{2}}\left(w-\frac{1}{w}\right)}\dfrac{dw}{w} + \dfrac{1}{4}\int_0^\infty e^{-\frac{i}{2}(x^2-t^2)^{\frac{1}{2}}\left(w-\frac{1}{w}\right)}\dfrac{dw}{w}\ldots.(66)$

In the first integral on the right of (66), the limits of w may be taken from 0 to $+i\infty$, so that the path of integration is the upper half of the imaginary axis. In the second integral, the limits may be taken to be 0 and $-i\infty$.

If we put $w = is$ in the first case, and $w = -is$ in the second,

$$E = \frac{1}{2}\int_0^\infty e^{-\frac{(x^2-t^2)^{\frac{1}{2}}}{2}\left(s+\frac{1}{s}\right)}\frac{ds}{s}$$

$$= \int_1^\infty e^{-\frac{(x^2-t^2)^{\frac{1}{2}}}{2}\left(s+\frac{1}{s}\right)}\frac{ds}{s}.$$

Or, if we write $\delta + \dfrac{1}{\delta} = 2\alpha$,

$$E = \int_1^\infty e^{-(x^2-t^2)^{\frac{1}{2}}\alpha}\frac{d\alpha}{(\alpha^2-1)^{\frac{1}{2}}} = f\left[(x^2-t^2)^{\frac{1}{2}}\right].\ldots\ldots(67)$$

When $t > x$, we substitute $u = w\,\dfrac{(t-x)^{\frac{1}{2}}}{(t+x)^{\frac{1}{2}}}$ in (65).

Then
$$E = \frac{1}{2}\int_0^\infty \cos\left[\frac{(t^2-x^2)^{\frac{1}{2}}}{2}\left(w+\frac{1}{w}\right)\right]\frac{dw}{w}$$
$$= \int_1^\infty \cos\left[\frac{(t^2-x^2)^{\frac{1}{2}}}{2}\left(w+\frac{1}{w}\right)\right]\frac{dw}{w},$$

or
$$E = \int_1^\infty \cos\left[(t^2-x^2)^{\frac{1}{2}}\alpha\right]\frac{d\alpha}{(\alpha^2-1)^{\frac{1}{2}}} = \frac{\pi}{2}K_0\left[(t^2-x^2)^{\frac{1}{2}}\right].\ldots(68)$$

The function $\frac{\pi}{2}K_0$ has been tabulated by Smith (see J. H. Michell, *Phil. Mag.*, Jan. 1898).

The curves show the early stages of the dispersion. Since $f(x_0)$ becomes infinite as $-\log x_0$ when x_0 is small, it appears that the electric force is infinite at the origin when $t=0$. It is infinite at $x=t$ at time t. But the total energy is finite since $\int_0^\infty [f(x_0)]^2\,dx$ is finite.

Reference to § 3 will explain Figs. 6 to 10. The curve gives the value of E at times 0, 1, 2, 3 and 5.

A wave of electric force moves out in each direction from the origin. The secondary disturbances due to the motion of the electrons cause the waves which appear between $x=t$ and $x=-t$. Since E is a function of x^2-t^2, the zero values of E are at points travelling outwards from the origin. The velocity of motion of a zero at x at time t is equal to $\frac{t}{x}$.

The effect of the secondary waves due to the motion of the electrons is to make E at first negative between $x=t$ and $x=-t$; as t increases, the sign of E becomes again positive near the origin. Since E is a function of x^2-t^2, the zero values of E all travel outwards from the origin. When t is large, the number of zeroes between $x=t$ and $x=-t$ is $\frac{2t}{\pi}$. For large values of t the zeroes of $K_0(t)$ are at intervals of π, and two zeroes move outwards one in each direction at any time t when $K_0(t)=0$.

Near the origin, when t is large, we have waves of very large wave-length for which $\lambda=0$. Hence when t is large and x is finite, the equation
$$\frac{d^2E}{dt^2} + E = \frac{d^2E}{dx^2}$$

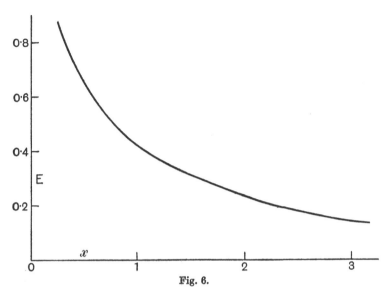

Fig. 6.

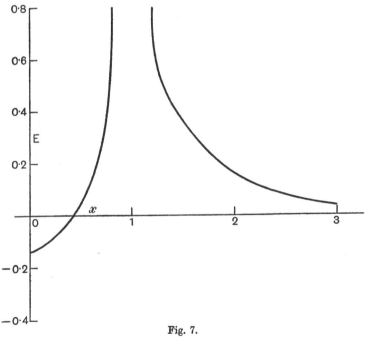

Fig. 7.

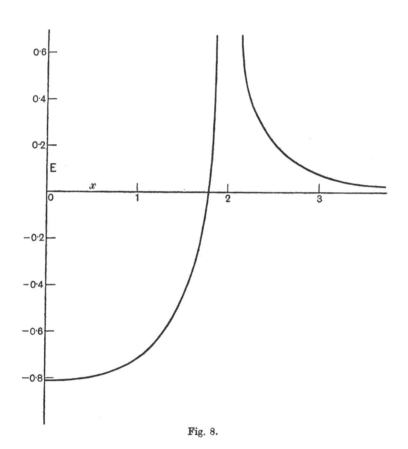

Fig. 8.

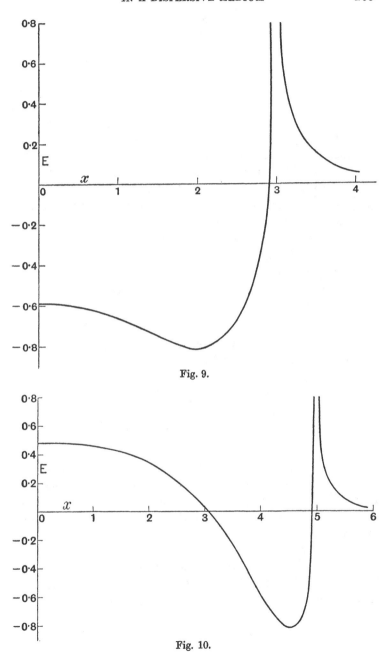

Fig. 9.

Fig. 10.

becomes
$$\frac{d^2 E}{dt^2} + E = 0, \quad \dots\dots\dots\dots\dots(69)$$

and E is zero at intervals of $\frac{1}{\pi}$.

The equation (69) indicates a mode of oscillation of the energy between the æther and the electrons moving in it which is not accompanied by a flux of energy outwards. It is from the region in which x is small and in which (69) holds approximately that the waves indicated by (68) all issue. Thus the process of dispersion presents two features.

In the region where the original disturbance arose, we have a system of standing waves. These arise because the electrons and the æther immediately surrounding them form a dynamical system having a period of vibration of its own. But the region to which the disturbance reaches increases with the time. The wave front moves with the velocity of light just as in free æther, and what would otherwise be merely standing waves are drawn out into a progressive wave train.

§ 7. The final distribution of energy may be deduced from (60) by the same method as that used in § 4.

Corresponding to (35) and (36), we have

$$v = \frac{1}{2\pi} \int_{-\infty}^{\infty} \frac{e^{i\delta t} - e^{-i\delta t}}{2i\delta} e^{i\lambda x} \, d\lambda \int_{-\infty}^{\infty} f(x_0) e^{-i\lambda x_0} \, dx_0, \quad \dots\dots(70)$$

$$H = -\frac{1}{2\pi} \int_{-\infty}^{\infty} \frac{e^{i\delta t} - e^{-i\delta t}}{2i\delta} e^{i\lambda x} \lambda \, d\lambda \int_{-\infty}^{\infty} f(x_0) e^{-i\lambda x_0} \, dx_0 \dots.(71)$$

These are deduced from (60) by using (23) and (24). The conditions $v = 0$ and $H = 0$ when $t = 0$ are satisfied. As in § 4, the final result will prove to be in agreement with that deduced from the Fourier analysis. In the medium we are now considering, an exceptional case arises when v or H is given initially and E is zero.

Suppose the initial conditions are that v is equal to v_0, a function of x, when $t = 0$, and that E and H are then zero. The equation (26) satisfied by v is not of the same form as (25). If we integrate (26) with respect to t, it gives

$$\frac{d^2 v}{dt^2} + v = \frac{d^2 v}{dx^2} + f(x), \quad \dots\dots\dots\dots(72)$$

$f(x)$ being a function of x only. Equation (22) is equivalent to $\dfrac{d^2v}{dt^2} + v = -\dfrac{dH}{dx}$, and since H is initially zero when $t = 0$, the left-hand side of (31) vanishes then.

It appears therefore that $f(x) = -\dfrac{d^2v_0}{dx^2}$, which is not zero except in a region where v_0 is initially zero, and (72) is not an equation of the same form as (25). E may be deduced directly from (25), $E = 0$ when $t = 0$, and (22) gives

$$\frac{dE}{dt} = -v_0 \text{ when } t = 0,$$

so that $E = -\dfrac{1}{2\pi} \displaystyle\int_{-\infty}^{\infty} \dfrac{e^{i\delta t} - e^{-i\delta t}}{2i\delta} e^{i\lambda x}\, d\lambda \int_{-\infty}^{\infty} v_0 e^{-i\lambda x_0}\, dx_0, \ \ldots(73)$

and the conditions as to E are satisfied.

The value of v may be got from (73) by using (24) and recalling that $v = v_0$ initially, whence

$$v = \frac{1}{2\pi} \int_{-\infty}^{\infty} \frac{e^{i\delta t} + e^{-i\delta t}}{2\delta^2} e^{i\lambda x}\, d\lambda \int_{-\infty}^{\infty} v_0 e^{-i\lambda x_0}\, dx_0$$

$$+ \frac{1}{2\pi} \int_{-\infty}^{\infty} \frac{\lambda^2}{\delta^2} e^{i\lambda x}\, d\lambda \int_{-\infty}^{\infty} v_0 e^{-i\lambda x_0}\, dx_0, \ldots(74)$$

where $\lambda^2 + 1 = \delta^2$, and when $t = 0$ in (74), the right-hand side becomes the Fourier expansion of v_0. In the second term of (74), we can perform the integration with respect to λ:

$$\frac{1}{2\pi} \int_{-\infty}^{\infty} \frac{\lambda^2}{\lambda^2 + 1} e^{i\lambda(x - x_0)}\, d\lambda = -\tfrac{1}{2} e^{-(x - x_0)}, \text{ if } x > x_0,$$

$$= -\tfrac{1}{2} e^{+(x - x_0)}, \text{ if } x_0 > x.$$

These results follow from a simple contour integration. A difficulty arises when $x = x_0$. We then perform the integration with respect to x_0 in (74) over a range from $x - \epsilon$ to $x + \epsilon$. The part of the second term of (74) contributed by the values of x_0 within this range is

$$\frac{v_0(x)}{2\pi} \int_{-\infty}^{\infty} \frac{\lambda^2}{\lambda^2 + 1} \frac{(e^{i\lambda\epsilon} - e^{-i\lambda\epsilon})}{2\lambda}\, d\lambda, \ \ldots\ldots\ldots\ldots(75)$$

$v_0(x)$ being the value of v_0 when $x_0 = x$.

As $\quad \dfrac{1}{2\pi}\displaystyle\int_{-\infty}^{\infty} \dfrac{\lambda^2}{(\lambda^2+1)}\dfrac{e^{i\lambda\epsilon}}{i\lambda} = \tfrac{1}{2}e^{-\epsilon} = \tfrac{1}{2}$ in the limit,

and $\quad \dfrac{1}{2\pi}\displaystyle\int_{-\infty}^{\infty} \dfrac{\lambda^2}{(\lambda^2+1)}\dfrac{\epsilon^{-i\lambda\epsilon}}{i\lambda} = -\tfrac{1}{2}e^{-\epsilon} = -\tfrac{1}{2}$ in the limit,

the expression (75) is therefore equal to $v_0(x)$, the value of v_0 at x.

Hence (74) becomes

$$v = v_0(x) - \frac{1}{2}\int_{-\infty}^{x} v_0\, e^{-(x-x_0)}\, dx_0 - \frac{1}{2}\int_{x}^{+\infty} v_0\, e^{(x-x_0)}\, dx_0$$
$$+ \frac{1}{2\pi}\int_{-\infty}^{\infty} \frac{e^{i\delta t} + e^{-i\delta t}}{2\delta^2}\, e^{i\lambda x}\, d\lambda \int_{-\infty}^{\infty} v_0\, e^{-i\lambda x_0}\, dx_0 \dots\!(76)$$

The last term of (76) represents the part of the disturbance which ultimately takes the wave form. But the first three terms of (76) are independent of the time, and so part of the energy is never actually analysed into waves.

Printed in the United States
By Bookmasters